ECONOMICS OF
AQUACULTURE

The Author

Dr. R.K.P. Singh is a Senior Scientist of Agricultural Economics at Rajendra Agricultural University, Pusa, Samastipur. He started his career as Senior Research Investigator in Administrative Staff College of India, Hyderabad. Before joining Rajendra Agricultural University, Pusa he worked as Marketing Specialist in Bihar State Agricultural Marketing Board, Patna. Dr. Singh has been teaching in almost all the faculties of the University namely; Agriculture, Dairy, Fisheries, Agricultural Engineering and Home Science. Dr. Singh has published more than 150 research papers and popular articles in reputed national and international journals. His book on Dairy Development Through Co-operatives got immense appreciation among the scientists and policy makers engaged in dairy development. He has conducted 10 research projects sponsored by the Ford Foundation, International Rice Research Institute, (Manila), Indian Council of Agricultural Research, New Delhi, University Grants Commission, New Delhi and Indian Council of Scientific Research, New Delih. More than two dozen post graduate students worked under his guidance for their research work. Dr. Singh headed the expert team of the University for implementing DWCRA project financed by UNICEF.

He is one of the IDA fellowship recipients at Ph.D. level. He has been awarded Visiting Associateship by the University Grants Commission, New Delhi for his excellence in Agricultural Economics Research.

ECONOMICS OF AQUACULTURE

R.K.P. Singh

Department of Agricultural Economic
Rajendra Agricultural University, Pusa
Samastipur (Bihar)

2013

DAYA PUBLISHING HOUSE

A Division of

Astral International Pvt. Ltd.

New Delhi - 110 002

© Author
First Published, 2002
Reprinted, 2013

ISBN 978-93-5124-147-8 (International Edition)

Published by	:	**Daya Publishing House®**
		A Division of
		Astral International Pvt. Ltd.
		– ISO 9001:2008 Certified Company –
		4760-61/23, Ansari Road, Darya Ganj
		New Delhi-110 002
		Ph. 011-43549197, 23278134
		E-mail: info@astralint.com
		Website: www.astralint.com
Laser Typesetting	:	**Classic Computer Services**
		Delhi - 110 035
Printed at	:	**Chawla Offset Printers**
		Delhi - 110 052

PRINTED IN INDIA

Acknowledgements

This book is based on the final report of an ad-hoc research project entitled "An Economic Analysis of Fish Production and Marketing in North Bihar" which was undertaken with the financial support provided under the cess fund by the Indian Council of Agricultural Research, New Delhi-1. I am grateful to the ICAR for their support.

Author owes debts of gratitude to Dr. S. R. Singh, Vice-Chancellor, Rajendra Agricultural University, Pusa for his encouragement during the period of preparation of final manuscript. I am also grateful to Dr. P.N. Jha, Former Vice-Chancellor & Dean (Agri.), Rajendra Agricultural University, Pusa, Samastipur for their constructive advice, encouragement and help in conducting this study.

I had very useful discussions with my colleagues in different stages of this study for which I am thankful to them. For their co-operation and intense interactions I am thankful to Dr. B.N. Verma, University Professor and Chairman, Dr. C.P. Yadav, Dr. R.N. Yadav, Assoc. Prof., Dr. B.B. Singh, Assoc. Prof., Dr. A.K. Choudhary, Asstt. Prof. and Dr. S.K. Jha, Asstt. Prof., Dr. L.N. Singh, Asstt. Prof., Deptt. of Agril. Economic, Rajendra Agril. University, Pusa.

I owe a deep sense of gratitude to my reverend teacher Dr. D.K. Singh, Retd. Professor of Agril. Economics and Dr. J.N. Choudhary, Retd. HOD, Agril. Economics, TCA, Dholi, Rajendra Agril. University, Pusa, Samastipur, Dr. D. Jha, National Professor (Agril. Economics), ICAR, National Centre for Agril. Economics Research and Policy, New Delhi, Sri Ramadhar (Retd. IAS) Member, the Commission on Agril. Costs and Prices, Ministry of Agri. and Co-operation, Govt. of India, Krishi Bhawan, New Delhi for their encouragement and blessing which made possible the completion of this work.

Respondent farmers, labours and officials of Fishery Co-operative Societies also deserve thanks for their co-operation during the field survey. Several Scientific personnel and students contributed to compilation of information, however the help of Mr. K.K. Prasad, Mr. Sanjeev Kumar Singh, Mr. Prafull Jha and Mr. Vinod Kumar is acknowledged for their assistance at the every stage of this research project.

We wish to express special thanks to Md. Nasim Mr. Arun Kumar, Md. Wazair, for their secretarial assistance. It is needless to add that all observations and conclusions are my own and, for any errors in interpretation and judgement still remain in this publication, I am alone responsible.

R.K.P. Singh

Contents

List of Tables

1

INTRODUCTION

Importance of Fish culture

There is an old proverb – "Give a man a fish and you feed him for a day, teach him how to fish and you feed him for a lifetime". But this proverb does not hold true in the present situation. As human population increases and natural fisheries resources diminish, knowing how to fish is not enough for today's fishers and their families. In the present dynamic economic situations, fishermen would be better off by learning how to grow fish or trying another trade altogether. Global fish supply is becoming increasingly scarce and more subject to human influences. The transition to relative scarcity cannot be prevented by more intensive fishing but rather will be ameliorated by better management of fisheries resources and improved acquaculture production.

It is a well know fact that fish farming has edge over crop production since it can be conducted on land that is not suited for crop production. It can flourish on land whose waters are mostly saline. Further, as fish live in a fluid medium and are cold blooded, they require minimal metabolic energy for maintenance of body temperature and for normal locomotion compared with land animals. Hence, it may be said that they are the most efficient converters of food. When various vertebrates are properly fed balanced diets under favourable environmental conditions, the conversion rates of dry feed to wet weight gain are as follow : fish about less than 1.5; cattle

---

Ronsivalli, L.H. (1976). "The role of Fish in meeting the world's Fish needs", Massive Fisheries Review, 39, No.6, 1-3, National Manure fisheries service, Seattle, Wash.

Bell, F. W. and E.R. Canterbery (1976). Aquaculture for the developing countries; A feasibility study, Cambridge, Mass Ballinger Publishing Co.

about 10.0 to 1.0, hogs, 4.0 to 1.0 and poultry 2.5 to 1.0 (Ronsivalli, 1976). Fish also use space more efficiently than many land animals because they are three dimensional habitants. In well managed environments, 3000 kg or more of fish can be produced per hectare per year, contrasting the maximum figure for cattle is 500 to 700 kg (Bell and Caterbery, 1976).

In addition a desired amount and quality of fish can be made available to consumers through fish culture. Fish farmers can also control production and market their stock when natural supplies are either seasonally low or unavailable for other reasons. Moreover, acquaculture offers the possibility for species improvement by selective breeding to meet consumer's tastes and market's requirements.

Acquaculture can also become a major income generating component in our Integrated Rural Development Programmes. It can be practised as supplementary enterprise to crop production and animal husbandry for generating employment and income to improve the quality of life of poor section of rural society. Further, culturing exportable species of fish would contribute to foreign exchange earnings. All of these directly or indirectly may help improving nutrition and employment, and consequently an increase in income on rural households.

India is the seventh largest producer of fish in the world and second in inland fish production. Fishery sector plays a vital role in sustaining a fairly large proportion of population particularly along the 8129 kms of coast line. The contribution of fisheries to the net domestic products has increased from Rs. 921 crore in 1984-85 to Rs. 10700 crore in 1998-99 at current price, showing about eleven and half fold increase during almost two decades. During the period, the share of fishery increased from 0.75 per cent to about 1 per cent of National Gross Domestic product. It has immense potential for export since the export of fish and fish preparation increased from Rs. 5 crore in 1960-61 to Rs. 5114 crore in 1999-2000 which accounted for 1.76 per cent and 2.08 per cent of total agricultural exports of the country, respectively. The quantum of fish export also increased from 19.9 thousand tonnes in the year 1960-61 to 390.6 thousand tonnes in the year 1999-2000 which accounted for 1.72 per cent and 6.90 per cent respectively of total fish production in the country (Appendix –I and Figure-1).

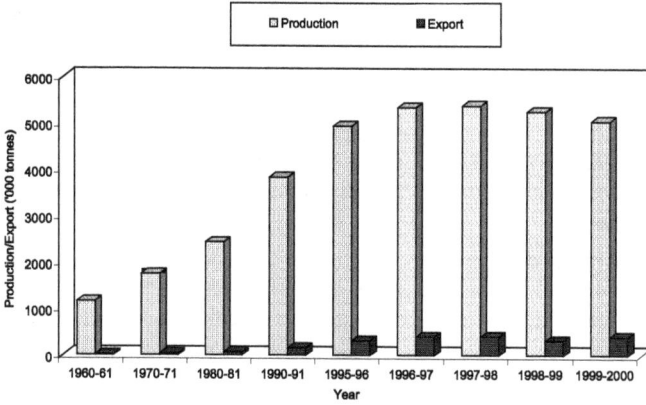

Fig. 1: Fish Production and Export from India During 1960-61 to 1999-2000.

Inland fish culture

Among the two types of fish production sources, the inland fishery has great advantage of production within the area of consumption or very close to area of consumption. It reduces the problems of preservation and transportation consequently the cost of distribution. The system of inland fish production can be classified on different bases also. An idea of various types of fish culture is necessary for assessment of its economic efficiency and comparative study. Interpreting economic results without considering specific conditions of a given fish culture can lead to incorrect generalisations. Hence, it is necessary to know the criteria for classification of acquaculture system in Bihar which could be understood through information given in Appendices IIA and IIB.

Fishery Resources

Bihar is land locked state hence inland fish could only be produced. There is about 4.08 lakh ha of water area which accounts for 2.35 per cent of geographical area of State (Table 1.1). An analysis of permanent and seasonal water areas in the state, revealed that the permanent water area is much higher (2.60 lakh ha) than seasonal water area (1.48 lakh ha) which account for 1.50 per cent and 0.85 per cent of geographical area, respectively.

An attempt has been made to discuss the zone-wise water spread since Bihar is divided in six homogenous agro-climatic zones on the basic of rainfall, temperature, humidity, soils, and aquifer (Map-I). Intensity of rainfall and type of soils have direct bearing on fish production. The high level of rainfall facilitates the fish production whereas fish may not be produced successfully in sandy loam soils because water permeability is high and water retention capacity of pond is low.

Among different six agro-climatic zones of the state, Zone I has comparatively higher water area of 1.18 lakh ha but the proportionate water area to the respective geographical area of the zone is comparatively higher in Zone II (5.15%). Zone I ranked first with respect to both permanent and seasonal water areas. Among plain region (new Bihar), Zone II is the smallest but it is placed on 2nd position in area under permanent water. Moreover, the comparatively higher proportion of permanent water area to geographical area was observed in Zone II (3.30 percent). Zone IV,

Table 1.1: Zone-wise geographical area, permanent and seasonal water area in Bihar

Zone	Geographical Area in lakh ha	Permanent Water Area		Seasonal Water Area		Total Water Area	
		Area in lakh ha	Per cent to the geo-graphical area	Area in lakh ha	Per cent to the geo-graphical area	Area in lakh ha	Per cent to geo-graphical area
I	38.37 (22.14)	0.80 (30.77)	2.08	0.38 (25.68)	0.99	1.18 (28.92)	3.08
II	17.86 (10.31)	0.59 (22.69)	3.30	0.33 (22.30)	1.85	0.92 (22.55)	5.15
III	45.00 (25.97)	0.58 (22.31)	1.29	0.48 (32.43)	1.07	1.06 (25.98)	2.36
IV	30.00 (17.31)	0.38 (14.62)	1.27	0.19 (12.84)	0.63	0.57 (13.97)	1.90
V	26.63 (15.36)	0.14 (5.38)	0.53	0.08 (5.41)	0.30	0.22 (5.39)	0.83
VI	15.44 (8.91)	0.11 (4.23)	0.71	0.02 (1.35)	0.13	0.13 (3.19)	0.84
Total	173.30 (100.00)	2.60 (100.00)	1.50	1.48 (100.00)	0.85	4.08 (100.00)	2.35

Figures in parentheses indicate the percentage of respective total of the state.

Zone V and Zone VI are placed in plateau region (now Jharkhand state) and constitute about 41.59 of the geographical area but have only 22.55 per cent of water area of the state. Bihar plains comprising Zone I, Zone II and Zone III (new Bihar) constitute 58.41 per cent of geographical area of undivided Bihar but 77.45 per cent of water area of the state is located in these zones. Water area is a necessary condition for fish production hence it may be said that there is a potential of fish production in Bihar plains, in general and north Bihar, in particular.

In future water area may increase due to increase in irrigation infrastructure and construction of ponds and reservoirs. A large number of ponds may also be constructed in future under various Govt. Programmes including Fish Farmers Development Agencies. About 10 per cent of 50 lakh ha of paddy land is also under deep water ecosystem which may be utilized for rice-fish cultivation in future.

In Bihar there are four major fisheries ecosystems like riverine fisheries ecosystem, reservoir fisheries ecosystem, semi-confined fisheries ecosystem and confined (ponds) fisheries ecosystem. Water area under different ecosystems are presented in Table 1.2.

Among four fisheries ecosystem, reservoir fisheries and pond fisheries have immense potentiality since modern fish technologies could be adopted in these areas. If fish productivity is increased upto 5000 kg/ha, the fish production in Bihar may reach to about one million tonnes from ponds and reservoirs only.

Table 1.2: Water area in different fisheries ecosystems in Bihar

Sl. No.	Eco-systems	Area (in lakh ha)
1.	Riverine fisheries system	1.78
2.	Reservoir fisheries system	0.97
3.	Semi-confined water fisheries system	0.38
4.	Confined (Pond) fisheries system	0.95
	Total water area*	4.08

*   Water area is excluding deep water rice area which could be utilized for rice fish culture.

Riverine fishery ecosystem

The net work of the various river systems in Bihar has a total length of about 3200 kms and constitutes the most important source of capture fishery. The river systems in the flood plains, particularly of the Ganga and its tributaries with vast stretches of canals, abandoned river meanderings, lateral pools and depressed lands during the rainy seasons are rendering a mosaic of natural fisheries with dynamic balancing on pressure of exploitation and conditions favouring their regenerations.

The river systems of Bihar can be categorised on the basis of three regions of Bihar.

(i)   Rivers of south Bihar

(ii) Rivers of North Bihar

(iii) Rivers of Chota Nagpur and Santhal Parganas

Ganga covers about 71,680 sq. kms and accounts for about 42% of the state river surface. The other rivers of north Bihar fall into the river Ganges at different places. The rivers of south Bihar and also the Chhotanagpur rivers fall into the river Ganges. The following are the major rivers of Bihar :-

North Bihar

(1) Koshi, (2) Kamala, (3) Mahananda, (4) Bagmati and Adhwara group, (5) Burhi Gandak, (6) Gandak, and (7) Ghagra

South Bihar

(1) Karamnasa, (2) Sone, (3) Punpun, (4) Kiul and Harchar, (5) Badua, and (6) Chandan

Jharkhand State (Chhotanagpur and Santhal Parganas)

(1) Swarnrekha, (2) Ajajana, (3) Damodar and (4) South Koel, and (5) Sankh

Reservoir fisheries ecosystem

Reservoirs that spread over an area of nearly one lakh hectares may play a pivotal role in increasing the fish production and help generating gainful employment to the poor section of rural society in the state of Bihar.

The reservoirs of Bihar may be categorised into three distinct categories *i.e.*, small, medium and large reservoirs, which are spread over to 12,461 ha (112 reservoirs), 12,523 ha (5 reservoirs) and 71,711 ha (8 reservoirs), respectively (Table 1.3). Palamu district of Chhotanagpur region (Jharkhand State) has the largest number of small reservoirs (58), however, large reservoirs are situated in the districts of Hazaribagh, Dhanbad, Snathal Pargana, Gumla, Ranchi and Giridih (one each).

Of the 125 reservoirs in the state, 112 (90%) are small and their water area is estimated to 13% of the total water area occupied by reservoirs in the state. The medium reservoirs also occupy almost identical water area but 74 per cent of the water spread is under only eight large reservoirs.

Table 1.3: District-wise number of major reservoirs and their area in Bihar

Districts	Small (less than 1000 ha)		Medium (1000 to 5000 ha)		Large (5000 ha and above)		Total	
	Number	Area (ha)	Number	Area (ha)	Number	Area (ha)	Number	Area (ha)
Santhal Parganas	-	-	1	3846	1	10000	2	13846
Garhwa	3	340	-	-	-	-	3	340
Bhagalpur	8	1164	2	2385	-	-	10	3549
Ranchi	19	3363	1	3500	1	16000	21	22863
Hazaribagh	3	243	1	2792	2	13209	6	16244
Giridih	-	-	-	-	1	6000	1	6000
Munger	8	1999	-	-	-	-	8	1999
Godda	1	257	-	-	-	-	1	257
Dhanbad	-	-	-	-	2	19002	2	19002
Lohardaga	1	305	-	-	-	-	1	305
Rohtas	1	513	-	-	-	-	1	513
Nawada	1	11	-	-	-	-	1	11
Singhbhum	2	161	-	-	-	-	1	161
Palamu	58	1994	-	-	-	-	58	1994
Gumla	7	2111	-	-	-	7500	8	9611
Total	112	12461	5	12523	8	71711	125	96695

Compiled from Ahmed and Singh (1992) and Supplemented by data provided by the Director of Fisheries, Govt. of Bihar.

Semi-confined water fisheries ecosystem

There are two types of semi confined water systems namely; ox-bow lakes and chaurs. Ox-bow lakes are mostly clustered in the East and West Champaran districts. These two districts together hold 41 ox-bow lakes with total normal water spread of about3500 hectares. There are a few more in Samastipur, Muzaffarpur and Sitamarhi districts. In the flood plains of Bihar, there are sprawling shallow inundated areas neither utilized properly for crop farming nor for fish culture with their amorphous outline, water coverage widely vacillates between wet and dry months which also depends on the intensity of rains in a particular year. Hence, it is difficult to estimate the precise area under this ecosystem. In Champaran (East and West), Darbhanga, Madhubani, Samastipur, Sitmarhi and Vaishali districts where chaurs and swamps are pre-dominant, the coverage is estimated to 35 thousand hectares.

Chaurs and swamps generally supported wild fishery of trash and air breathing species.

Pond fisheries ecosystem

It has been estimated that there are about 60,000 tanks and ponds in the state, covering total water area of over 95,000 ha. Zone wise number of ponds/tanks and their water area are presented in Table 1.4.

It may be observed from the table that the number and water area of Govt. ponds were comparatively higher (30187 and 70.81 thousand ha, respectively) than that of the private ponds (28953 and 24.31 thousand ha, respectively). Zone-wise analysis revealed that the number and water area of Govt. ponds were comparatively higher in Zone I that is; 10125 and 24.08 thousand ha, respectively. Except in Zone VI, the water area of Govt. ponds was comparatively higher in all the zones than the water area of private ponds. Bihar plains comprising Zone I, Zone II and Zone III, constituted about 68 per cent of pond's water area of the state. However, 36 per cent of ponds and 46per cent of pond's water area of the undivided state are located in north Bihar that is; Zone I and Zone II. However, these two zones (north Bihar) have about 67 per cent of pond's water area of divided Bihar. Hence, it may be said that there is a quite substantial water resource available in north Bihar for fish production.

Table 1.4:   Zone-wise number and area of ponds/tanks in Bihar

(Area in 000'ha)

Zone	Govt. Ponds		Private Ponds		All Ponds	
	Number	Water area	Number	Water area	Number	Water area
I	10125 (33.54)	24.08 (34.01)	5154 (17.80)	6.09 (25.07)	15279 (25.83)	30.17 (31.72)
II	1621 (5.37)	9.87 (13.94)	4550 (15.71)	3.46 (14.23)	6171 (10.43)	13.33 (14.01)
III	5446 (18.04)	16.81 (23.74)	2912 (10.06)	4.71 (19.36)	8358 (14.13)	21.42 (22.52)
IV	9858 (32.66)	9.92 (14.01)	78.01 (26.94)	5.11 (21.03)	17659 (29.86)	15.03 (15.80)
V	1478 (4.89)	7.97 (11.26)	3524 (12.17)	2.31 (9.49)	5002 (8.46)	10.28 (10.81)
VI	1659 (5.50)	2.16 (3.04)	5012 (17.31)	2.63 (10.81)	6671 (11.34)	4.78 (5.03)
Total	30187 (100.00)	70.81 (100.00)	28953 (100.00)	24.31 (100.00)	59140 (100.00)	95.01 (100.00)

Figures in parentheses indicate the percentage of respective State totals.

Fish Production

In undivided Bihar, total fish production from different fishery ecosystems increased continuously from 46.40 thousand tonnes at the triennium ending 1960-61 to 254.7 thousand tonnes at the triennium ending 1999-2000 and it is placed at third position in inland fish production in the country after West Bengal (8.66 lakh tonnes) and Andhra Pradesh (3.80 lakh tonnes) in the year 1999-2000. Despite an increase in fish production, state's share in national inland fish production declined from 17.40 per cent at the triennium ending 1960-61 to 9.01 per cent at the triennium ending 1999-2000 (Table 1.5).

It is worth pointing out that the inland fish production increased by more than 9 fold at country level during the period 1960-61 to 1999-2000 whereas the corresponding increase was only 5 fold in the state of Bihar. As far as state's share in total fish production (Inland + Marine) at country level is concerned it remained almost constant (about 4.50 per cent) during last 2 and half decades, probably due to comparatively slow growth in marine fish production in the country (23.00 lakh tonnes in 1990-91 to 28.34 lakh tonnes in 1999-2000).

Table 1.5: Fish production in Bihar *vis-a-vis* India

(in 000 tonnes)

Year (Triennium ending)	India			Bihar	
	Inland	Marine	Total	Production	% share in National Fish production
1960-61	266.00	781.66	1047.66	46.4 (17.4)	4.43
1970-71	661.66	967.33	1628.99	55.3 (8.36)	3.39
1980-81	887.00	1555.5	2442.5	94.5 (10.66)	3.87
1990-91	1431.00	2130.6	3561.6	155.5 (10.86)	4.36
1995-96	2110.3	2682.6	4792.9	211.8 (10.04)	4.42
1996-97	2206.3	2752.0	4958.3	228.9 (10.35)	4.59
1997-98	2353.6	2874.7	5228.3	234.4 (9.96)	4.48
1998-99	2565.8	2696.4	5262.2	203.3 (7.88)	3.84
1999-2000	2827.7	2833.9	5656.6	254.7 (9.01)	4.50

Figures in parentheses indicate the percentage of inland fish production in Bihar to inland fish production at country level.

Source : Tabulated on the basis of data published in Agricultural Statistics At A Glance, 16th Series (2001), Department of Economics and Statistics, Ministry of Agriculture and Co-operation, Govt. of India, New Delhi p.203.

Zone wise analysis of fish production revealed that the Bihar plains constituting Zone I, Zone II and Zone III contributes about 65.47 per cent of total fish produced in the state of Bihar. Plateau region constituting Zone IV, Zone V and Zone VI is spread over to 45 per cent of the geographical area of the state but contributes only 34.53 per cent of total fish production (Table 1.6).

Among the six zones, fish production has been comparatively higher in Zone I (74.64 thousand tonnes) accounting for about 30 per cent of state's fish production, followed by Zone III (23.30 percent), Zone IV (19.10 per cent), Zone II (12.29 per cent) Zone V (10.57 per cent) and Zone VI (4.86 per cent). During nineties (upto 1996-97), all the zones of Bihar plains and Zone IV performed comparatively better in fish production by improving their proportionate share in state fish production whereas fish production in Zone V declined from 28.34 thousand tonnes in 1989-90 to 26.40 thousand tonnes in 1996-97. Fish production in Zone VI increased from 7.75 thousand tonnes in 1989-90 to 12.15 thousand tonnes in 1996-97 but it's share in state's fish production declined from 14.95 per cent in 1989-90 to 4.80 per cent in 1996-97.

Table 1.6: Zone-wise fish production in Bihar during 1989-90 to 1996-97

(in tonnes)

Zones	1989-90	1990-91	1991-92	1992-93	1993-94	1994-95	1995-96	1996-97
I	45606	47284	58600	47770	48860	53210	71360	74640
	(29.13)	(29.56)	(31.69)	(29.17)	(24.96)	(27.23)	(29.78)	(29.88)
II	16785	17465	19860	19300	26760	30660	29480	30700
	(10.72)	(10.92)	(10.74)	(11.76)	(13.67)	(15.69)	(12.30)	(12.29)
III	35322	36914	42332	37647	48010	47250	53930	58190
	(22.56)	(23.08)	(22.89)	(22.95)	(2453)	(24.18)	(22.51)	(23.30)
IV	22752	24301	30738	26940	36820	37821	36720	47700
	(14.43)	(15.19)	(16.62)	(16.42)	(18.81)	(19.36)	(15.33)	(19.10)
V	28335	26917	24560	22410	28260	16950	39340	26400
	(18.10)	(16.83)	(13.28)	(13.66)	(14.44)	(8.67)	(16.42)	(10.57)
VI	7750	70.49	8800	10000	7000	9500	8750	12150
	(4.95)	(4.41)	(4.76)	(6.10)	(3.58)	(4.88)	(3.65)	(4.86)
Total	156559	159930	184890	164067	195710	195390	239580	249780
	(100.00)	(100.00)	(100.00)	(100.00)	(100.00)	(100.00)	(100.00)	(100.00)

Figures in parentheses indicate the percentage production of the state.

Source : Department of Fisheries, Govt. of Bihar, Patna

Zone-wise annual compound growth rates in fish production were estimated which are presented in Table 1.7.

An analysis of growth in fish production in Bihar revealed that it has increased by 6.71 per cent per annum during 1989-97 which was found statistically significant at 1 per cent level of probability and much higher than the growth rate achieved in agriculture sector in Bihar. Zone-wise analysis revealed comparatively higher growth in fish production in Zone II (10.09%), followed by Zone IV (9.74%), Zone III (7.10%), Zone I (6.24%), Zone VI (4.88%) and Zone V (0.06%) during 1989-97. Annual growth rates in fish production in all the zones, except Zone V were statistically significant.

Table 1.7: Zone-wise annual compound growth rate in fish production during 1989-97

Zone	Growth Rate	t-Calculated Value
I	6.24	3.220**
II	10.09	7.160*
III	7.10	7.104*
IV	9.74	6.856*
V	0.62	0.157
VI	4.88	2.046***
Bihar	6.71	6.070*

*   Significant at 1 per cent level of probability.
**   Significant at 5 per cent level of probability.
***   Significant at 10 per cent level of probability.

Rationale

The above discussions clearly indicates that the increase in fish production has been faster than the increase in production under different sectors of the state economy. Despite the increase in fish production from 0.46 lakh tonnes in 1960-61 to 2.02 lakh tonnes in the year 1998-99, the State performance in inland fish production has been inferior to many states of the country, particularly in late nineties. Its share in national inland fish production has declined from 17 per cent in 1960-61 to 9.02 per cent in the year 1998-99. In Bihar, fish productivity has been quite low in riverine reservoir and semi-confined ecosystem since generally capture fisheries are being practised in these ecosystem. Fish culture is practised in ponds but the per hectare fish productivity varies for 350 kg to 2800 kg in Bihar (Singh and Prasad, 2000). Hence, there is a scope for increasing fish production in the state of Bihar since per hectare yield potential of 3000 kg is quite common without much inputs on farmer's ponds.

There are two important dimensions of fish production in Bihar, that is predominance of Government owned fish ponds and fishery co-operative societies which are expected to exert much influence on fish production system in the state. Besides fish marketing system is also expected to play an important role in increasing the profitability to fish farmers which would ultimately motivate them to invest more as well as to adopt improved technologies in fish farming to increase fish production. Bihar is, no doubt, still a deficit state in fish production and a large quantity of fish is supplied through different states of the country. The process of fish marketing which is still most disorganised, has also not attracted effective attention of institutional agencies.

There is an urgent need to make institutional efforts in the field of research, extension and training to exploit the fish production potentiality which have been in utter negligence in the state of Bihar. There is hardly any comprehensive research in the field of fish culture, in general and economics of fish production and marketing, in particular conducted in Bihar situation. Hence, the present research project has been planned to study the economics of fish production and marketing system in the state of Bihar. The detailed objectives of the study are :

Objectives of the study

(1) To study the common practices of fish production in different regions of North Bihar.

(2) To estimate the cost, production and profitability in fish culture.

(3) To examine the extent and pattern of employment in fish production.

(4) To understand marketing practices, market intermediaries and channel of fish marketing.

(5) To identify the major constraints in fish production and marketing with the help of case studies of Fishery Co-operatives.

Layout of the Project Report

Project report is presented in ten chapters. The first chapter presents an Introduction with objectives of the investigation. Methodological approach adopted for conducting the research has been presented in the second chapter. This is followed by the chapter of findings and discussion which includes profile of fish farmers and ponds as third chapter, cultural practices as fourth chapter, cost of fish production as fifth chapter, fish production, potential production, production efficiency and profitability as sixth chapter. The extent and pattern of employment has been discussed in the seventh chapter and fish Marketing is examined in detail in eighth chapter. The ninth chapter deals with constraints in fish production. The tenth and last chapter contains a brief summary, conclusions, and emerging issues for increasing fish production which followed by Bibliography and Appendices.

2
METHODOLOGICAL APPROACH

The study is based on primary data obtained through field survey from representative respondents. The project was conducted in North-Bihar since 46 per cent of fish produced in un-divided Bihar and 67 per cent of divided Bihar is grown in north Bihar only. There are two agro-climatic zones in north-Bihar which are known as Zone -I (North-West Alluvial Plains) and Zone II (North- East Alluvial Plains). Zone-I compresses 12 districts and Zone II comprises 9 districts (Appendix-III)

Identification of districts

It was not practicable to draw samples from each districts of north Bihar hence a representative sample of 6 districts i.e. three districts in each of Zone I and Zone II were identified purposively for drawing samples of blocks. The care was taken to select districts which were covered under the project of Fish Farmers Development Agency and have comparatively larger water area and quantum of fish production. Zone-wise names of districts under study alongwith their water area and fish production are presented in Table 2.1.

Table 2.1: Water area and fish production in identified districts

Zone/District	Water Area (ha)	Production at Triennium ending 1997 (in tonnes)
Zone-I		
Samastipur	1386.13	6255
Darbhanga	3036.50	8166
East Champaran	4003.54	6186
Zone-II		
Madhepura	1509.17	4663
Purenea	3447.85	5240
Katihar	4175.89	5973

Methodological Approach

A list of block alongwith pond's area was prepared for all the six districts which were identified for drawing representative samples of blocks, villages and fish farmers. In each districts, blocks were arranged in descending order on the basis of their water area and three blocks in each identified districts with comparatively larger water area, making a sample of 18 blocks were identified for drawing sample of villages. The names of selected blocks in each identified districts are presented in Table 2.2.

Table 2.2: Names of selected Blocks alongwith their water area

Selected Blocks	Water Area (ha)
(A) District E. Champaran	
1. Dhaka	257.8
2. Ghorasahan	204.0
3. Chiraiya	214.0
(B) District Darbhanga	
4. Singhwara	258.00
5. Keoti	225.60
6. Manigachhi	154.00
(C) District Samastipur	
7. Hasanpur	305.00
8. Singhia	290.00
9. Samastipur	82.00
(D) District Katihar	
10. Kadwa	362.62
11. Azam Nagar	332.88
12. Manihari	311.90
(E) District Purnea	
13. Kirtyanand Nagar	169.58
14. Amaur	152.32
15. Baisi	126.96
(F) District Madhepura	
16. Chausa	113.17
17. Puraini	98.57
18. Madhepura	87.65

Selection of Nucleus villages

Names of villages of identified blocks alongwith their number of fish ponds and water area were obtained either from the respective

Block Development Office or the concerned District Fisheries Office. A separate list of villages with their number of fish ponds and water area was prepared for each sample blocks. Village with the largest number of ponds/water area in a particular block was identified as sample nucleus village. A sample of 18 nucleus villages that is; one nucleus village from each identified block were selected for drawing sample of fish farmers and fish labours.

In case of less than 10 fish farmers in the identified nucleus village, two to four adjacent villages were identified to make a cluster of villages to have sufficient number of fish farmers for drawing 10 sample respondents. Names of block-wise identified villages (Nucleus and associate villages) alongwith number of fish farmers, fish ponds, and water area are presented in Table –2.3.

Table 2.3: Block-wise sample villages alongwith number of fish farmers, ponds and water area

Sl. No. Block	Sample Village	No. of fish Farmers	No. of Ponds	Water Area (ha)
1. Chiraya	1. Gangapiper*	5	6	6.2
	2. Mishrauliga	8	8	5.8
	3. Rampur	6	7	4.4
	4. Semera	8	8	5.3
2. Ghorashan	1. Kadamwa*	9	9	6.4
	2. Dhanupra	8	8	5.8
	3. Jagirha	6	6	5.0
	4. Unarpur	7	7	4.8
3. Dhaka	1. Bararwa-Lakhansen*	6	9	9.6
	2. Chamhi	7	7	6.0
	3. Ahiraulia	7	7	6.6
	4. Gondri	7	7	7.0
4. Manigachhi	1. Kaisthkobari*	10	13	10.0
	2. Narayanpur	9	10	7.6
	3. Raghopur	8	8	5.8
5. Keoti	1. Keoti*	8	11	8.4
	2. Ranbe	8	8	6.8
	3. Paigmberpur	6	7	6.2
6. Singhwara	1. Bhajaura*	15	19	12.8
	2. Kaligaon	11	12	8.0
	3. Asthua	10	10	6.4

contd....

Table 2.3 contd...

Sl. No.	Block	Sample Village	No. of fish Farmers	No. of Ponds	Water Area (ha)
7.	Hasanpur	1. Rampur-Rajwa*	6	6	5.80
		2. Patepur	3	3	5.84
		3. Malhipur	2	2	1.07
8.	Singhia	1. Morwara*	7	10	5.72
		2. Kaina	2	2	0.84
		3. Khairpura	2	3	1.32
9.	Samastipur	1. Korbadhha*	4	5	10.06
		2. Mohanpur	1	1	0.88
		3. Bhawanpur	2	2	0.76
		4. Magardahi	2	2	7.07
		5. Lagunia-Raghkan	4	5	7.07
10.	Kadwa	1. Pagwa*	7	7	3.57
		2. Khanua	4	6	3.10
		3. Bharti	4	5	2.10
11.	Azamnagar	1. Brahmpur*	6	9	4.31
		2. Balua	5	5	2.61
		3. Rathi	4	4	1.85
12.	Manihari	1. Sapta*	7	7	3.46
		2. Suhasan	5	5	2.58
		3. Bhatwara	4	4	1.80
13.	Kirtyanand Nagar	1. Baghmara*	7	9	4.46
		2. Bela	6	6	2.98
		3. Haishali	4	5	2.25
14.	Amaur	1. Daripur*	6	8	3.98
		2. Pothiaya	4	5	2.57
		3. Kohabari	4	4	1.89
15.	Baisi	1. Aura*	7	8	3.93
		2. Dhadhaur	5	5	2.40
		3. Harna	4	5	2.44
16.	Chiraiya	1. Paina*	8	10	4.98
		2. Dhaneshpur	7	7	3.19
		3. Morsand	5	6	2.95
17.	Puraini	1. Karma*	8	10	5.16
		2. Balia	1	2	0.79
		3. Kursandi	3	4	2.20
18.	Madhepura	1. Batauna*	9	10	4.85
		2. Sakarpura	5	5	2.48
		3. Parariya	4	4	2.15

*   indicate the nucleus village.

Selection fish farmers

A list of fish farmers with the number and area of ponds operated by them was prepared separately for all 18 clusters of villages. Fish farmers were arranged in descending order on the basis of their water area (pond's area)

A sample of 10 fish farmers in each identified villages, making the sample size of 180 fish farmers were selected randomly with the help of Random Table. Selected farmers were categorised in three groups on the basis of size of the ponds that is; small ponds owners (less than 0.50 ha), medium pond owners, (0.5 to 1.0 ha) and Big pond owners (1.0 ha and above). The similar exercise was done for all the 18 clusters of villages. Pond size group wise number of sample farmers, number of ponds and water area are presented in Table 2.4.

Table 2.4: Size group-wise sample farmers alongwith their ponds and water area

Category of Ponds	No. of Sample farmers	No. of Sample ponds	Water Area (ha)
Small	68 (37.78)	76 (37.62)	20.23 (14.85)
Medium	69 (38.33)	79 (39.11)	52.00 (38.17)
Large	43 (23.89)	47 (23.27)	63.99 (46.98)
Total	180 (100.00)	202 (100.00)	136.22 (100.00)

Figures in the parentheses indicate percentage to the respective totals.

Selection of Fishing Workers

One main fishing worker, working either as attached labour or worked for larger number of days on the ponds of identified fish farmers formed the sample for fishing workers. Hence, there were 180 fishing workers who formed the sample of fishing workers in the study.

Selection of fishery co-operative society

All the selected blocks had one fishery co-operative society. At first, all the 18 fishery co-operative societies were selected for detailed investigation but the societies officials did not provide any information since these societies were engaged in only making arrangements for leasing in ponds from the Government. Almost all the Fisheries Co-operative Societies failed to provide any supporting services to the member farmers, except leasing of ponds.

Selection of Market Intermediaries

All the district level Regulated Agricultural Markets of six sample districts, that is Samastipur, Darbhanga, East Champaran, Katihar, Purnea and Madhepura formed the sample of markets for this study. The list of wholesalers, vendors and retailers who were dealing with fish marketing was prepared separately for these three market functionnaries for all the six Regulated Agricultural Markets. A sample of 3 wholesalers, 3 vendors and 3 retailers from each sample Regulated Agricultural Markets, making a sample of 18 wholesalers, 18 vendors and 18 retailers were selected randomly for detailed investigation. Moreover, all the sample 180 fish-farmers also formed the sample for the study of fish marketing practices and price spread at pond / village level.

Period of study

All the primary data obtained from fish farmers, fishery workers, co-operative societies, marketing intermediaries refer to the period 1998-99.

Method of Enquiry

The survey method of enquiry was followed for data collection. Data were collected from respondents through specifically prepared schedules for the study. There are 4 schedules which were prepared for fish-farmers, fishing workers, market intermediaries and co-operative officials. Formats were also developed for collection of data relating to information of villages under investigation. All the schedules were tested and modified for non-responsive questions / items.

Respondents were interviewed by Research Associates and field investigators however the random checking was done by Principal Investigator.

Analytical Procedure

Data obtained from fish farmers were computed farmer/pond wise and tables were prepared on the basis pond categories like small, medium and big. Information relating to fish workers were also tabulated pond-size wise however the working hours were converted into adult male unit-equivalent to 8 hrs of work.

Analytical framework

Tabular analysis was done to reach at relevant conclusions. In order to study the association between out put and various inputs used, Cobb-Duglas production function was used. The following model of Cobb-Duglas production function was used.

$$Y = a.x_1^{b1} \times x_2^{b2} \times x_3^{b3} \times x_4^{b4} \times x_5^{b5} \times x_6^{b6} \times e^u$$

Where,

Y	=	Value of gross fish production (Rs./ha)
a	=	Intercept (constant)
x_1	=	Manure (Rs./ha)
x_2	=	Fertilizer (Rs./ha)
x_3	=	Lime (Rs./ha)
x_4	=	Fingerlings (Rs./ha)
x_5	=	Feed (Rs./ha)
x_6	=	Human labour (Rs./ha)
e^u	=	Error term
$b_1 - b_6$	=	regression coefficients of respective variables.

To examine the productivity of different inputs used in fish production, marginal value productivities of inputs were estimated by the following formula :

$$MVP\ x_i = \frac{bi\ Y}{Xi}$$

Where,

X_i	=	i^{th} input (i=1, 2..........6)
Y	=	Geometric mean of gross fish production
x_i	=	Geometric mean of ith input and
b_i	=	Elasticity of production of ith input

Evaluation and apportionment of Costs

Rent

It refers to payment made by fishermen to the Government for using pond for fish production. The rent is usually fixed on per pond basis by auction. Auction is done for normally a period of one to three years but the payment is made on yearly basis before stocking in the pond.

In case of own pond(s) imputed value of rent was charged which was determined on the basis of rent of similar ponds in the village.

There were some ponds in the project area which were owned by landlords/rural institutions. As many as 62 sample fish farmers leased-in ponds either from land lords or local institutions. Hence, actual payment made by fish farmers to owners of the ponds was charged as rent of the pond.

Depreciation

Depreciation charge was imputed by using the purchase price/cost of constructing, making or producing inventory and then depreciated it by applying straight line method. The following rates were used in estimating the depreciation of various items. In case of items which are utilised in one year, the total value of cost is considered as cost in the particular year.

Table 2.5: Items and rates of depriciation

Sl. No.	Items	Rate of Depreciate (in per cent)
1.	Net	25
2.	Spade	33
3.	Hut for chankidary	50
4.	Bucket	30
5.	Basket	50
6.	Knopiya	33

Interest

It refers to payment for the use of capital whether fixed or operating. It was inputed at the rate of prevailing average bank rate that is; 12 per cent per annum. In case of operating capital, interest was charged for six months but it was charged for a year in case of fixed capital. Actual interest rate was changed when capital purchased through loans, obtained particularly through non-institutional sources.

Hired labour

Generally skilled labours were hired for fish production in the project area. Actual wages paid to them were charged as cost of hired labour. They were generally paid in cash how ever they were provided with either light breakfast or food. In case of strenuous and important jobs like; stocking and harvesting, both breakfast and food were provided to them. The value of breakfast and food has been added to the cash wage of hired labours. The common wage of hired labour was Rs. 30 which varied from Rs. 30 to 40 per day for 8 hours of work. Average wage of female labour varied from Rs. 25 to 30 whereas per day average wage of child labour was Rs.20 only.

Attached labour

Fish farmers who belong to other than fisherman community (Mallah) or large pond owners generally engage skilled male labour on monthly or share basis. Monthly labours were charged on the basis of actual payment to them for performing activities related to fish production.

Family labour

Family labour was evaluated at the wage rate paid to hired labour, that is Rs. 30- the common wage paid to hired labours for 8 hours of work. Fish production is strenuous and tiring job which is generally done in the pond's water by adult males. Female and children were also engaged in fish production but they performed jobs which were done in out side ponds like, marketing and transportation of feeds, manures and fertilizers from farm house to ponds. The wages paid to female and child labours were charged on the basis of payment made to hired female/child labour for similar type of job. However, their working hours were converted to adult man unit on the basis of wages paid to them.

Material inputs

Material inputs include fertiliser, lime, toxicant, feed and manure. These inputs were evaluated on the basis of actual payment made for acquiring these inputs plus transportation cost. The wages of human labours engaged in these purchases were not included in the cost of these material cost. However, home produced manure was evaluated on the basis of market price.

Various cost concepts used

Estimates of cost of fish production on different cost concepts:

Cost A_1 =   All actual expenses in cash and kind incurred in production by owner including rent paid for leased in law.

Cost A_2 = Cost A_1 + Imported value of own land

Cost B = A_2 + interest on value of owned capital

Cost C =   Cost B + imputed value of family labour

Marketing cost

Marketing cost includes value of packaging materials, transportation cost, profit margins of different market functionaries and storage cost.

1. Packaging material includes bag, basket, rope, ice which were evaluated at their value of actual payment made by fish farmers for acquiring these materials. In case of home prepared basket, market price was charged.

2. Transportation cost was also charged as per actual payment made by fish farmers. When farmers used their own carriage/tractor, the market rate was charged.

3. Retailer's share was obtained by deducting the value of fish purchased from the wholesaler/producers and other related expenses from the value of the fish sold to the consumer.

4. Wholesaler's share was obtained by deducting the value of fish purchased from producers/traders and related expenses from the value of fish sold to retailers/vendors/consumers.

5. Producer's share was obtained by deducting the wholesaler's/retailer's/vender's marketing cost, margin and expenses incurred by the producers for marketing the produce from the price of consumers.

Measures of profits

Gross income

It is total value of main and by-product which is calculated by dividing the quantum of produce by market price.

Net income

Gross income *minus* total expenses of production cost of fish seeds, manures, fertilisers, lime, *Mahua* Cake, supplementary feed, wages of hired labour, and imputed value of unpaid family labour, depreciation, rent, interest on owned and working capital and marketing cost.

Family labour income

Gross income *minus* total expenses of production excluding wages of the unapid family labour.

Farm business income

Gross income *minus* total expenses of production excluding wages of family labour and interest on working capital. It is a measure of the earnings of a farmer and his family for their capital investment, labour and managerial work.

3

FINDINGS AND DISCUSSION

As mentioned in Chapter 1, the primary data were used to examine the different facets of fish production which have been analysed in detail and presented as findings and discussion in this chapter. At first, the profile of fish farmers has been discussed in the third chapter to get an idea about fish farmers' socio-economic level who are engaged in fish production in north Bihar. Pond is one of the necessary factors for fish production hence an attempt has been made to examine the area, water area, depth of water, ownership, leasing arrangement of sample ponds under investigation which are presented in the last section of the third chapter. The cultural practices of fish production has also examined in fourth chapter to know the level of input use and adoption extent of scientific fish production technology. The cost of fish production has been analysed in detail in the fifth chapter which includes cost estimates, use of inputs, fish seeds including sources of fish seeds. Fish production, conversion ratio, potential fish production, production efficiency and profitability in fish production have been discussed in the sixth chapter. The extent and pattern of employment of fishery labours have also been studied to know the generation of employment in rural area through fish production. The fish marketing is an important aspect of any economic activities since it has been discussed in the eighth chapter of the project report. The detailed discussions are presented in the text in above mentioned order.

Profile of Fish farmers

In this section of the study, the main thrust is to examine the socio-economic status of fish-farmers in north-Bihar. Keeping in view this objective, age structure, education status, family size, occupational structure, land ownership, possession of ponds and co-operative membership have been discussed in brief.

Age-structure

Fish farmers under study were grouped in three age categories that is; young farmers (below 30 years), matured farmers (30 –50 years) and old farmers (50 years and above) for three specified categories of ponds which are presented in Table 3.1.

Table 3.1: Age-wise distribution of fish farmers operating different categories of ponds

(Number)

Age-Group	Small Pond Operators	Medium Pond Operators	Large Pond Operators	All Ponds Operators
Young farmers (below 30 years)	6 (8.82)	5 (7.25)	2 (4.65)	13 (7.22)
Matured farmers (30-50 years)	51 (75.00)	48 (69.56)	31 (72.09)	130 (72.22)
Old farmers (50 years & above)	11 (16.18)	16 (23.19)	10 (23.26)	37 (20.56)
Total	68 (100.00)	69 (100.00)	43 (100.00)	180 (100.00)

Figures in parentheses indicate percentage to the respective totals.

It may be observed from the table that about three-fourths (72.22 per cent) of fish farmers belonged to matured age category and 20.56 per cent belonged to old age category. Only 7.72 percent of fish farmers were young in the project area. Almost the similar trend was observed for the three specified categories of pond owners however the comparatively lesser proportion of young farmers (4.65 per cent) and larger proportion of old farmers were found operating large size ponds. It was probably due to the fact that the large pond owners were comparatively economically superior to ponds owners of other categories and their young family members might have not preferred to work in fish culture. Moreover, the large pond owners had more of managerial work in fish production which could be done by old farmers, even in better way than young farmers due to their long experience in the field. It may further be observed that the comparatively less proportion of small pond owners belonged to old age category (16.18 per cent). It was mainly due to poor health status of small pond owners who were generally toiling below poverty line and withdrew themselves from work due to more strenuous jobs in fish culture.

Educational Status

Educational level of fish farmers has been examined since it affects the knowledge level, skill development, exposure to production technology and marketing practices and adoption level of improved technology. To assess the educational level of fish farmers under investigation, they were categorised in four groups that is; illiterate, below secondary, secondary and above secondary which are presented in Table 3.2.

Table 3.2: Educational level of fish farmers operating different categories of ponds

(Number)

Educational Level	Small Pond Operators	Medium Pond Operators	Large Pond Operators	All Pond Operators
Illiterate	28 (41.18)	23 (33.33)	17 (16.28)	58 (32.22)
Literate				
Below secondary	31 (45.59)	34 (49.28)	21 (48.83)	86 (47.78)
Secondary	6 (8.82)	78 (10.14)	9 (20.93)	22 (12.22)
Above Secondary	3 (4.41)	5 (7.25)	6 (13.96)	14 (7.78)
Total	68 (100.00)	69 (100.00)	43 (100.00)	180 (100.00)

Figures in the parentheses indicate percentage to the respective totals.

It may be observed from the table that about 32.22 per cent of fish farmers under investigation were illiterate and 67.78 per cent were literate. Rate of illiteracy increases with the decline in the size of the pond whereas just reverse trend was observed in case of literacy level that is; the higher the pond size, the higher is the educational level of fish farmers. In educated category 34.99 per cent of big pond owners belonged to education level of secondary and above whereas 17.39 per cent of medium pond owners and 13.23 per cent of small pond owners got education upto this level. It clearly indicates that the educational achievement is directly linked with size of the ponds that is, the economic status of fish farmers.

Family Size

The family size has been examined to have an idea about the supply of human labours for fishing activities. The comparatively

larger size of family is supposed to take care of fishing activities
through family labour. Fish farmers were categorised in three family
size groups that is, small family (< 5 members), medium family (5 –
8 members) and large family (8 and above members) which are
presented in Table 3.3.

Table 3.3:   Pond category-wise family size of fish farmers

(Number)

Family Size	Small Pond Operators	Medium Pond Operators	Large Pond Operators	All Ponds Operators
Small family (< 5 members)	19 (27.94)	18 (26.09)	12 (27.91)	49 (27.22)
Medium family (5 – 8 members)	34 (50.00)	38 (55.07)	15 (34.88)	87 (48.33)
Large family (8 and above members)	15 (22.06)	13 (18.84)	16 (37.21)	44 (24.45)
Total	68 (100.00)	69 (100.00)	43 (100.00)	180 (100.00)

Figures in parentheses indicate percentage to the respective totals.

It may be observed from the table that about 27 per cent of fish
farmers had small family, 48.33 per cent had medium family and
24.45 per cent had large family. It may be noted that all the categorise
of pond owners had almost identical proportion of small size of
family that is 26.09 to 27.94 per cent.

It may further be observed that the majority of families of small
and medium pond owners belonged to medium family size whereas
the comparatively large proportion of large pond operators (37.21%)
had large family size. The trend is in order of other studies. The
large pond operators have generally larger size of family because
the assets in general, and land in particular, keep family members
united whereas poor households have meagre asset and family
members disintegrate and prefer to live independently since these
families do not have asset which could attract them to live together
in joint family.

Caste structure

Caste is one of the important factors affecting the choice of the
occupation, and possession of skill in different rural economic
activities and it holds more true in Bihar because the state experienced

Profiles of Fish Farmers and Ponds

slow change in rural economy. Keeping this in view, the caste structure of fish farmers under study has been categories in four caste groups that is, upper caste, fishing caste (Mallah), other OBC, and lower cast which are presented in Table 3.4.

Table 3.4: Caste-categories of fish farmers operating different size of ponds

(Number)

Educational Level	Small Pond Operators	Medium Pond Operators	Large Pond Operators	All Ponds Operators
Upper Castes	-	6 (8.70)	5 (11.63)	11 (6.11)
Middle Caste				
(i) Fishing Castes	51 (75.00)	45 (65.22)	25 (58.14)	121 (67.22)
(ii) Other OBCs	9 (13.23)	13 (18.84)	12 (27.91)	34 (18.89)
Lower Caste	8 (11.77)	5 (7.24)	1 (2.32)	14 (7.78)
All Castes	68 (100.00)	69 (100.00)	43 (100.00)	180 (100.00)

Figures in the parentheses indicate percentage to the respective totals.

Table 3.4 revealed that about two-thirds of fish farmers (67.22 per cent) belonged to fishing caste (Mallah), 18.89 per cent to OBCs, 7.78 per cent to lower castes and only 6.11 percent to upper castes category. None of the upper caste fish farmers had small size of pond and only one lower caste fish-farmer had large size pond. Proportion of fish farmer belonging to upper caste and OBC category increases with the increase in the size of ponds where as the proportion of fishing caste (Mallah) and lower cast decreases with the increase in the size of ponds. It clearly indicates that the fish farmers of upper castes and OBC category are operating comparatively larger size of ponds than that of fishing caste and lower caste fish farmers. Moreover, the fishing caste (Mallah) has major stake in fish farming in north- Bihar.

Occupational structure

All the fish farmers under study were, no doubt, engaged in fish production but all of them did not have fish farming as main occupation. Respondent farmers had varied type of occupations either as main or subsidiary. All the occupations of respondent farmers were categorised in five groups that is; fish farming, crop farming, dairying, business and 'others'. Moreover, the majority of respondent farmers had two and more occupations. The occupation which generated either comparatively more employment or more income was considered as main occupation and the second important occupation in the order was considered as secondary occupation. The occupation which was third important, was not considered in the present study. Primary and secondary occupations of small, medium and large pond owners were computed which are presented in Table 3.5.

It may be observed from the table that the majority of respondent fish-farmers (76.67 per cent) had fish farming as main occupation and 21.11 per cent had fish farming as secondary occupation however the comparatively larger proportion of small pond owner (89.71 per cent) had fish farming as main occupation than the medium pond owners (76.81 per cent) and large pond owners (55.81 per cent). It has further been observed that the small pond owners had fish farming either as main or secondary occupation. It clearly indicates their dependence on fish farming. On the other hand, one medium pond owner and three large pond owners had fish farming as tertiary occupation. Crop farming was the second important main occupation for respondent fish farmers since 12.22 per cent of them had crop farming as main occupation. The importance of crop farming as main and secondary occupation increased with the increase the size of the pond of respondent fish farmers. Dairying was not important main occupation on sample households but it was practised by 25.56 per cent of respondents. Its importance as secondary occupation increased with the decline in the size of pond of respondents. Business and 'other' occupations were not important activities in the project area.

Size of Land Holdings

The size of land holding is one of the important indicators of social status in rural area. It also indicates resource base of farmers. The comparatively higher land base indicates higher level of resources and vice-versa. Respondent fish-farmers were grouped in three categories that is; land less (no land), small (< 2 ha) and large (2 ha and above). Number of fish farmers in each categories alongwith their net operated area have been computed which are presented in Table 3.6.

Table 3.5: Main and subsidiary occupation of fish farmers operating different categories at ponds

(Number)

Occupation	Small Pond Operators		Medium Pond Operators		Large Pond Operators		All Pond Operators	
	Primary Occupation	Secondary Occupation	Primary Occupation	Secondary Occupation	Primary Occupation	Secondary Occupation	Primary Occupation	Secondary Occupation
Fishing	61 (89.71)	7 (10.30)	53 (76.81)	15 (21.74)	24 (55.81)	16 (37.21)	138 (76.67)	38 (21.11)
Crop Farming	3 (4.41)	8 (11.76)	7 (10.14)	18 (26.09)	12 (27.91)	17 (39.53)	22 (12.22)	43 (23.89)
Dairying	1 (1.47)	21 (30.88)	3 (4.35)	20 (28.98)	2 (4.65)	5 (11.63)	6 (3.33)	46 (25.56)
Business	3 (4.41)	6 (8.82)	5 (7.25)	8 (11.59)	3 (6.98)	2 (4.65)	11 (6.11)	16 (8.89)
Others	-	11 (16.18)	1 (1.45)	3 (4.35)	2 (4.65)	1 (2.33)	3 (1.67)	15 (8.33)
All	68 (100.00)	53 (77.94)	69 (100.00)	64 (92.75)	43 (100.00)	41 (95.35)	180 (100.00)	158 (87.78)

Figures in parentheses indicate percentage to the total number of fish farmers under respective category of pond size group.

Table 3.6: Size of land holdings of fish farmers operating different size of ponds

(Area in ha)

Class of Household	Small		Medium		Large		All Ponds	
	No. of Farmers	Net Operated Area	No. of Farmers	Net Operated Area	No. of Farmers	Net Operated Area	No. of Farmers	Net Operated Area
Landless (No land)	47 (69.12)	-	21 (30.43)	-	5 (11.63)	-	73 (40.56)	-
Small (0-2 ha)	18 (26.47)	12.85 (46.04)	31 (44.93)	26.70 (16.86)	22 (51.16)	15.42 (9.74)	71 (39.44)	54.97 (15.95)
Large (2 ha above)	3 (4.41)	15.06 (53.96)	17 (24.64)	131.62 (83.14)	16 (37.21)	142.96 (90.26)	36 (20.00)	289.64 (84.05)
All	68 (100.00)	27.91 (100.00)	69 (100.00)	158.32 (100.00)	43 (100.00)	158.38 (100.00)	180 (100.00)	344.61 (100.00)

Figures in the parentheses indicate percentage to the respective totals.

It may be observed from the table that about 40.56 per cent of respondent fish farmers were landless who had no cultivable land. About 39.44 per cent fish farmers had land holding of less than 2 ha, and 20 per cent fish farmers had land holding of 2 ha and above. It may further be observed that about 11.63 per cent of large pond others, 30.43 per cent of medium pond owners and 69.12 per cent of small pond owners were landless. It may be pointed out that the landless house holds had large size of ponds because these farmers leased in ponds for fish–farming from the Government. It may further be observed that the proportion of small and large size land holdings increased with the increase in the pond size, indicating the larger pond owners had comparatively larger size of land holding.

Average size of land holding of respondent fish farmers was estimated to 1.91 ha however the ownership of land was skewed since 20 per cent of respondents had 84.05 per cent of total operated area and 39.44 per cent of respondents had only 15.95 per cent operated area owned by all the sample fish farmers under study. The skewedness in land distribution was more pronounced among large size pond owners than medium and small pond owners.

Hence, it may be inferred that the fish farming is an occupation of poor households because about 41 per cent fish farmers had no cultivable land and 39 per cent farmers had cultivable land which worked out to be less than 0.8 ha per household.

Co-operative membership

At first, Bihar, Fisheries Co-operative Societies Act was enacted in1935 which was revised in November, 1960. The organization and functioning of fisheries co-operative are based on provisions made in the Act and by-laws of the fisheries co-operative. The main objectives of fisheries co-operative are to inculcate the habit of thrift, self-reliance and extension of co-operative philosophy among fisherman community which are to be achieved through following activities.

1. To have right of fishing through contract or any other means.
2. To develop infrastructure for fish processing.
3. To arrange marketing of fish and fish products.

4. To arrange credit through institutional agencies or saving of co-operative for fish production and marketing.

5. To encourage fishermen for production and marketing of makhana and water nuts.

6. To hire or construct building and other permanent assets to facilitate fish production and marketing.

In rural area, membership in any rural Institution is a symbol of social status. Membership of fishery co-operative society is not only a status symbol but it is necessary factor for getting Government fish pond on lease basis since Governments ponds are being leased out through fishery co-operative society in Bihar. All the sample fish farmers were grouped in two categories that is; member and non-member which are presented in Table 3.7.

Table 3.7: Distribution of fish farmers according to FCS membership on different categories of ponds

(Number)

Particulars	Small	Medium	Large	All Ponds
Member Farmers	38	44	25	107
	(55.88)	(63.77)	(58.14)	(59.44)
Non-member Farmers	30	25	18	73
	(44.12)	(36.23)	(41.86)	(40.56)
Total	68	69	43	180
	(100.00)	(100.00)	(100.00)	(100.00)

Figures in the parentheses indicate percentage to the respective totals.

It may be observed from the table that 59.44 per cent of sample fish farmer were member of fishery co-operatives and remaining 40.56 per cent did not have member ship fishery co-operative. The membership of fishery co-operative was comparatively higher in case of medium size pond owners (63.73 per cent) followed by large pond owners (58.14 per cent) and small pond owners (55.88 per cent). It clearly indicates that the medium pond owners were more aware of facilities of fishery co-operative society. Moreover, fishery co-operatives did not provide any services to member-farmers, except arranging the leasing of ponds. The low level of membership of fishery co-operatives might have the result of their poor performance in Bihar.

Table 3.8: Pond area, water area and range of water depth on different categories of sample ponds

Category of Ponds	Number of Ponds	Pond Area (ha)	Average size of Pond (ha)	Water Area (ha)	Proportion of Water area to pond area (%)	Range of Water depth (meter)		
						Monsoon	Winter	Summer
Small (< 0.50 ha)	76 (37.62)	22.65 (15.17)	0.30	20.23	89.31	1.40 to 4.60	1.00 to 3.40	0.00 to 2.00
Medium (0.5 to 1 ha)	79 (39.11)	54.45 (38.47)	0.73	52.00	90.52	2.2 to 6.50	1.5 to 4.80	0.00 to 3.00
Large (1 ha & above)	47 (23.27)	69.23 (46.36)	1.47	63.99	92.43	2.6 to 7.00	1.8 to 5.10	0.90 to 2.90
All Ponds	202 (100.00)	149.33 (100.00)	0.74	136.22	91.22	1.40 to 7.00	1.00 to 5.10	0.00 to 3.00

Figures in the parentheses indicate percentage to respective total.

Pond's Profile

In present section, an effort has been made to discuss the pond's area, water area, water depth, ownership, extent of ownership and source of lease in ponds.]

Pond's size and water depth

Size group-wise number of ponds, pond's area, average pond size and their average water depth are computed which are presented in Table 3.8.

It may be observed from the table that out of 202 ponds under study, 79 ponds (39.11%) were medium size (0.50 to 1.00 ha), 76 ponds (37.62%) were small (< 0.50 ha) and 47 ponds (23.27%) were large (1 ha and above) which where spread over the 54.45 ha, 22.65 ha and 69.23 ha, respectively. The large ponds constituted 23.27 per cent of total ponds under investigation but accounted for 46.36 per cent of area of all the sample ponds whereas the small ponds constituted 37.62 per cent area of sample ponds under investigation and accounted for only 15.17 per cent of area of total sample ponds.

The average size of ponds is worked out to be 0.74 ha however the average size of large ponds was 1.47 ha, average size of medium ponds was 0.73 ha, and of small pond 0.30 ha. The range of area of fishponds varied from 0.18 hectare to 0.48 hectare of small pond, 0.52 hectare to 0.88 hectares on medium ponds and 1 hectare to 5 hectares of large ponds.

It may further be observed from the table that the water area of ponds, on an average, was estimated to 91.22 per cent of total pond's area however it was comparatively higher on large ponds (92.43 per cent) followed by medium ponds (90.52 per cent) and small ponds (89.31 per cent). While analysing the water depth of fish ponds under investigation it was found that almost all the small ponds had no water in summer season however depth of water in these ponds varied from1.40 m to 4.60 m in monsoon and 1.00 m to 3.30 m in winter season. The depth increased with the increase in the pond's size that is, 2.20 m to 6.5 m and 2.60 m to 7.0 m in medium and big ponds respectively during monsoon, and 1.50 m to 4.80 m and 1.40 m to 5.10 m respectively in winter season.

The above discussions clearly indicate that the water depth in all the categories of ponds is sufficient for fish culture however medium and large ponds are most suitable for composite culture.

Pond's ownership and proprietorship

Size group wise ownership and extent of proprietorship of ponds were computed and presented in Table 3.9.

In the project are fish ponds were either owned or leased in through different sources which were operated either as sole proprietorship or as partnership. It may be observed from the table that about 80.69 per cent of sample ponds were leased in through different sources and 19.31 per cent ponds were owned by farmers themselves. Proportion of leased in ponds did not vary with the size of ponds.

Table 3.9: Size group-wise ownership and extent of proprietorship of ponds

(Number)

Category of Ponds	No. of Ponds	Ownership		Extent of Proprietorship		
		Owned	Leased in	Sole	Only 2 Partners	More than 2 Partners
Small	76	13	63	70	6	-
	(100.00)	(17.10)	(82.90)	(92.11)	(7.89)	
Medium	79	18	61	61	15	3
	(100.00)	(22.78)	(77.22)	(77.21)	(18.99)	(3.80)
Large	47	8	39	30	4	13
	(100.00)	(17.02)	(82.98)	(63.83)	(8.51)	(27.66)
Total	202	39	163	161	25	16
	(100.00)	(l9.31)	(80.69)	(79.70)	(12.38)	(7.92)

Figures in the parentheses indicate percentage to the respective totals.

It may further be observed from the table that the sole proprietorship was common form of fishery enterprise in the project area because 79.70 per cent of fish ponds were operated under this system however the comparatively large number of small ponds (92.11 per cent) were operated under sole proprietorship followed by medium ponds (77.21 per cent) and large ponds (63.83 per cent). Out of 202 ponds, 25 ponds were operated by two partners and 16 ponds were operated by more than 2 partners. It may be noted that none of the small ponds were operated by more than two partners whereas 13 large ponds (27.66 per cent) were operated by more than two partners.

Hence, it may be said that the smaller ponds are generally operated under sole proprietorship and the comparatively large number of medium and big ponds were operated under partnership arrangement.

Leasing source and Duration

As discussed in the preceding section, more than 80 per cent of ponds were leased in for fish production in the project area. These ponds were leased in through three sources, that is private, public and others (Panchayat, Schools, Trust etc.) for different durations which varied from 1 year to10 years.

These leased in ponds were computed for specified three sources and for the three time periods (1 year, 2-3 years and 4-10 years) which are presented in Table 3.10.

Table 3.10: Size group-wise leased in ponds alongwith their leasing source and lease duration

(Number)

Category of Ponds	Lease in Ponds	Leasing Source			Lease Duration		
		Private	Public	Others	1 Year	2-3 Years	4-10 Years
Small	63 (100.00)	22 (34.92)	34 (53.97)	7 (11.11)	41 (65.08)	13 (20.63)	9 (14.29)
Medium	61 (100.00)	6 (9.84)	42 (68.85)	13 (21.31)	48 (78.69)	5 (13.11)	(8.20)
Large	39 (100.00)	11 (28.21)	25 (64.10)	3 (7.69)	28 (71.79)	(28.21)	–
All Ponds	163 (100.00)	39 (23.93)	101 (61.96)	23 (14.11)	117 (71.78)	32 (19.63)	14 (8.59)

Figures in the parentheses indicate percentage to the leased in ponds in respective category of ponds.

It may be observed from table that the majority of ponds (61.96 per cent) were leased in from public (Government) source 23.96 per cent were leased in private (landlord) source and 14.11 per cent from 'other' sources. The main source of leasing in ponds for fish production was government in the project area however the private source was second important source for leasing small (34.92 per cent) and large (28.21 per cent) ponds whereas other source was second important source of leasing of medium ponds in the project area.

While analysing lease durations of fish ponds, it has been

observed that the majority of ponds (71.78 per cent) were leased in for one year, 19.63 per cent of ponds for 2-3 years and only 8.59 per cent ponds for 4 – 10 years. Pond size wise analysis of leasing durations revealed that the comparatively large proportion of medium ponds (78.69%) were leased for one year followed by big ponds (71.79 per cent) and small ponds (65.08 per cent). More than one fourth of big ponds (28.21 per cent) and one fifth of small ponds (20.63 per cent) were leased in for 2-3 years period and only 13.11 per cent of medium ponds were leased in for this period. None of the big ponds were leased in for more than 3 years whereas 14.29 per cent of small ponds were leased in for 4- 10 years.

Hence, it may be inferred that the majority of ponds were leased in from Government and the common period of leasing ponds was one year however the large ponds were leased in for the shorter time period and *vise-versa.*

4

CULTURAL PRACTICES

Fish production is a traditional enterprise in the project area. Earlier it was practised with traditional method but some scientific inputs and practices have been introduced during last twenty years mainly after the launching the project of Fish Farmers Development Agency in almost all districts of north Bihar. In the present section, an effort has been made to study cultural practices adopted by farmers in fish production.

Fish cultural practices include all activities from pre-stocking operation to harvesting of fish through human efforts. Prevailing cultural practices in the project area have been categorised on the basis major operations done by fish farmers like pre-stocking, stocking, input application, application of supplementary feeds and harvesting.

Pre-stocking

It refers to operations prior to stocking for the purpose of cleaning the pond water by weeding, netting and other manual activities because presence of unwanted materials like plant weeds, insects, frogs and predatory fish affect adversely the level of fish production. Hence, it is necessary to eradicate them from ponds to facilitate a better environment for proper growth and movement of fingerlings. Generally it was practised from the last week of May to mid of August. However, this operation is not practised in ponds were stocking was done just after harvesting, particularly in late harvested ponds. Out of 202 ponds under study, pre-stocking operation was done in 24 ponds only. About 17 human labour days were engaged in this operation for conducting this operation in one hectare pond. It is worth mentioning that none of the sample farmers reported about the use of weedicide for eradication of plant weeds

from the pond and it was performed by only manual labours in the
project area.

Stocking

Stocking refers to application of external fish seeds into the
pond for the purpose of raising them upto marketable size. Generally
it was practised from 2nd week of June to mid of September, however
only 13, out of 202 ponds were stocked in last week of September to
1st week of October. The peak stocking period was July in the project
area. There were two important types of fish culture prevailing in
the project area which may be termed as Indigenous major carp
culture and exotic carp culture. Moreover, the majority of ponds
were stocked by composite carp culture, i.e. indigenous and exotic.
Indigenous major carp culture includes fish species namely; Rohu,
Catla, Mrigal, minor carps and cat fishes. Whereas exotic carp culture
includes culture of exotic species namely; Common Carp, Silver Carp
and Grass Carp. As fish seeds is an indispensable input in fish
culture, all the 202 sample ponds were stocked by indigenous major
carps with or without exotic carps. However, 14.36 per cent of sample
ponds were also stocked by a specific type of indigenous fish species
i.e. Mangur. Among the exotic fish species common carp was used
on 35.15 per cent ponds, silver carp on 21.78 per cent ponds and
grass carp on 30.69 per cent of sample ponds. The indigenous major
carps i.e. Rohu, Catla and Mrigal are still most common fish species
for stocking in north Bihar since their proportion in total quantity of
stocking has been estimated to two-thirds of total fish seeds used in
north Bihar. The indigenous species mangur was introduced
recently in the project area and getting popularity in north Bihar,
particularly in East Champaran district, mainly due to comparatively
higher price and demand for this species of fish. However, on an
average, stocking proportion of Mangur was about 48 per cent only.
The proportionate stocking of exotic species i.e. common carp, silver
carp and grass carp was estimated to 22.53 per cent. The size of
indigenous major carps varied from 1 inch to 2.5 inches and that of
mangur 1.5 inch to 2.5 inches, whereas, the size of exotic species
ranged from 1.5 to 3.5 inches in the project area (Appendix-IV).

Among the indigenous species, Rohu was the most common
species as it was stocked in all the sample ponds and its share in
total quantity of fish seeds was more than 35 per cent whereas,
common carp was found most common species among exotic carps
of fish as its proportion of stocking quantity was observed to be

more than 40 per cent. The stocking proportion of different spices i.e. Rohu, Catla, Marigal, Common Carp, Silver carp, Grass Carp and Mangur were 27.93, 25.75, 20.31, 9.18, 5.03, 8.32 and 3.48, respectively. Human labours required for stocking in one hectare pond was estimated to 11.30 mandays in the project area. While field survey respondents were asked about the proper proportion of stocking of different species in the pond. It is worth pointing out that almost all the sample farmers showed their complete unawareness about the proper stocking proportion, indicating the poor fishery extension services in the project area.

Input Application
Lime

Fish thrives well in alkaline water. The application of lime in ponds is only to increase and/or maintain the alkaline nature of fish ponds. Unhydrous powdered lime is sprayed over the pond water for the purpose of maintaining the alkaline nature of pond water. Moreover it also helps in purifying the water. Fish farmers reported that it facilitates hygienic environment in ponds for protecting fish from water borne diseases. It was pratised generally before the stocking operation in the project area. About 23 percent of sample ponds were also sprayed by lime after stocking operation upto the month of December. As many as 75.74 per cent of sample ponds were sprayed by lime however wide variation in per hectare use of unhydrous lime (50 kg to 1600 kg) indicated poor knowledge of farmers about the recommended dose of lime for fish production.

Toxicants

Toxicants are the substances which are used to eradicate predatory fish, insects and frogs from the pond. The most common toxicant was *Mahua* oil cake which was sprayed over the pond water in powdered form, however it was sprayed 2 to 5 weeks before stocking in the project area. Toxicants are rarely used in fish production in north Bihar as only 4 ponds, out of 202 sample ponds were treated by the toxicant. It is worth mentioning that none of the small ponds were treated by toxicant in the project area, however only one pond of medium size group and three ponds of large size group were treated by the toxicant. Per hectare quantity of mahua oil cake utilized as toxicant in ponds ranged from 817.32 kg to 1762.71 kg against recommendation of 2000-2500 kg for scientific fish production. Hence, it maybe said that the use of toxicant was not only uncommon in the project area but per hectare use of toxicant (Mahua cake) was much lower than the recommendation.

Manures

Manure is not directly consumed by fish but it enhances the vegetative growth of green aquatic plants and phytoplanktons which are ultimately consumed by the phytophagous fish. Dung was the most commonly used manure in north Bihar which was added in as many as 55 per cent of the sample ponds, however the method and time of its application varied widely from pond to pond in north Bihar. Dung was applied in majority of ponds during August to December at interval of 1 month whereas it was applied in 38.74 per cent of the pond in single dose at the time of stocking. As far as method of application is concerned, dung was applied in 18.92 per cent of ponds by mounting the dung at the corner of pond however the majority of ponds (78.38 per cent) were sprayed by dung either over the whole pond water or all the sides of pond water, indicating the over utilization of manpower in the manuring. Hence it may be said the use of manure is still uncommon and fish farmers used it at very low level. Fish farmers are required to be educated for application of right quantity of manures at right time/interval in ponds.

Fertilizers

The purpose of adding fertilizer in the pond is almost similar to that of manure. It also facilitates the growth of macro and micro vegetative plants for an ultimate consumption by the phytophagous fish. The most common fertilizers used in the project area were urea and single super phosphate, whereas a few farmers also used di-ammonium phosphate with or without mixing potash. Fertilizers were sprayed over the pond water in one to 3 split doses on 28.22 per cent of sample ponds during October to January. Fish farmers prefer to use fertilizers at the time of low temperature perhaps knowing the fact of nitrogen loss in high temperature and such farmers also practised application of urea in split doses upto thrice instead of a single dose. Out of 57 fertilized ponds, as many as 40.35 per cent of the ponds were incorporated by urea and SSP in two to three split doses, 31.58 per cent of the ponds were fertilized by DAP only in a single dose, and 21.05 per cent of the ponds were incorporated by Potash with Urea, SSP or DAP in the project area. The farmers are yet to be aware of recommended dose of fertilizer and combination thereof. On enquiry about the knowledge of fertilizers use with respect to recommended dose of inorganic fertilizers for fish production, they expressed their unawareness in this matter. It may be due to poor institutional effort to acquaint them regarding the benefits of fertilizer use in fish production in north Bihar. Moreover, the low level use of fertilisers and chemicals are necessary for production of exportable fish since exportable fish should not carry any trace of fertilizers and chemicals

Supplementary feeds

Supplementary feed is a critical input in fish production and being used for the purpose of better and fast growth of fish. It becomes indispensable when natural feeds are not sufficiently available in the pond. Mustard oil cake and rice bran were most common supplementary feeds in north Bihar. Out of the 202, 7 ponds were applied by groundnut oil cake. Supplementary feeds were sprayed along the sides of the pond water during November to February in North Bihar at one day interval to weekly interval. None of the sample households reported daily application of supplementary feeds, probably due to the fact that regular application of feeds require heavy monetary expenses. As far as recommended dose of supplementary feeds is concern, the respondents showed their complete unawareness with this regard. It clearly indicated that supplementary feeds are applied into the ponds, no doubt, for better growth of fish but by hit and trial method only in the project area. It was observed that a total of 48 ponds, out of 202 sample ponds, were applied by supplementary feeds and out of 48 ponds, as high as 64.58 per cent of the ponds were incorporated by mixture of mustard oil cake and rice bran, and 6.25 per cent of the ponds were incorporated by mixture of Groundnut oil cake and rice bran. The mixture of mustard oil cake, rice bran and groundnut oil cake were applied only on 14.58, 6.25 and 8.33 per cent of sample ponds respectively.

Fish Harvesting

Harvesting is the last cultural practice in fish production. In north Bihar, it is spread over almost whole of the year but mainly concentrated during the months of December–May, however number of harvesting ranged from 3 to 6. About 45 per cent of ponds were harvested during March–May, but about 70 per cent of large ponds were harvested in April–May. Harvesting during February- March was quite common in north Bihar but it is more true for small and medium ponds. Large pond owners preferred to harvest fish in slack agricultural season. About two-thirds of the large ponds were harvested within two–three days, since they require a large quantity offish so that it could be sold in district market. About 74 per cent of small pond owners and 48 per cent of medium pond owners started harvesting in the later part of December to meet their family, agricultural and fish production expenses however they used to harvest only bigger size of fish. Moreover, early harvesting was also done in fear of poaching because it is very common in fish ponds, particularly owned/operated by weaker section.

It may be noted that none of small fish ponds were harvested in the month of April – May because harvesting on these ponds was usually completed latest by the month of March due to fear of drying of ponds. Contrary to this none of large ponds owners started fish harvesting before February.

It has further been observed that the harvesting operation is the domain of hired labour in north Bihar. Fish harvesting was done by mainly of hired labour (84 per cent), and only 16 per cent of family labours were employed for this operation in study villages. The proportion of family human labour employment for fish harvesting was comparatively higher on small size of ponds (26.20%) than medium size ponds (16.92%) and large size ponds (4.93 per cent). However, this finding is just reverse the findings of other studies conducted at country level (Gupta 1984). Moreover, there may also some change in employment pattern of labour use in fish production during last 16 years in other parts of country.

5

COST OF FISH PRODUCTION

In the present section the cost of fish production has been analysed in detail. The expenses incurred in fish production are categorised in two groups; that is fixed and variable costs. Fixed cost includes inputed rental value/rent of the ponds, interest and depreciation of fixed assets like net, spade, bucket, lantern, topiya, torch, structure on ponds. Variable cost includes human labour expenses on fingerlings, manures, fertilizers, toxicants, lime, feed, packaging materials, marketing cost etc. Per ha variable expenses and fixed expenditures for three specified categories of fish ponds were computed which are presented in Table 5.1.

It may be observed from the table that the total per ha, on an average, cost of fish production was Rs. 19.06 thousand, constituting 73.71 per cent variable and 26.29 per cent fixed costs. The comparatively higher per ha cost of fish production was observed on medium size of ponds (Rs. 21.56 thousand) followed by small ponds (Rs. 21.46 thousand) and large ponds (Rs. 16.28 thousand). Per ha variable cost had the similar trend. But per ha fixed cost was also comparatively higher on small ponds (Rs. 5.90 thousand) which declined with the increase in the pond size that is; Rs. 5.55 thousand on medium ponds and Rs. 4.29 thousand on big ponds. However, the proportion of fixed cost to total cost did not keep the similar trend since the proportion of fixed expenditure was comparatively lower on medium size of ponds (25.76 per cent) and higher on small size of ponds (27.47 per cent). Moreover, the comparatively per ha lower fixed cost was incurred in fish production on big size of ponds. It is expected also because fixed cost gets distributed on larger area of ponds.

Per ha variable cost was comparatively higher in fish production on medium size of ponds (Rs. 16.00 thousand) followed by small

Table 5.1: Per hectare variable and fixed costs incurred in fish production on different categories of ponds

(in Rs.)

Particulars	Small Cost	Small Proportion to total cost	Medium Cost	Medium Proportion to total cost	Large Cost	Large Proportion to total cost	Average Cost	Average Proportion to total cost
(A) Variable Expense								
1. Manures	283.49	1.32	276.35	1.28	163.38	1.00	224.34	1.18
2. Fertilizers	138.16	0.64	200.19	0.93	182.53	1.12	182.68	0.96
3. Toxicants	–	–	52.73	0.24	337.08	2.07	178.48	0.94
4. Lime	990.36	4.61	1530.77	7.10	1174.95	7.22	1283.36	6.73
5. Fingerlings	4578.35	21.33	4292.46	19.91	3230.11	19.84	3835.88	20.12
6. Supplementary feed	120.61	0.56	149.33	0.69	100.87	0.62	122.30	0.64
7. Human labour	7235.59	33.71	5975.73	27.72	3405.28	20.92	5513.27	28.92
8. Packaging	158.48	0.74	317.83	1.47	462.54	2.84	360.23	1.89
9. Marketing cost including transportation	304.35	1.42	624.58	2.90	801.22	4.92	660.00	3.46
10. Misc. Expenses	981.51	4.57	1456.00	6.75	1162.99	7.15	1247.89	6.55
11. Interest on working capital	776.52	3.62	1128.09	5.23	964.33	5.92	998.95	5.24
Total Operating Cost (A)	15567.42	72.53	16004.06	74.24	11985.28	73.64	14051.37	73.71
(B) Fixed Expenses								
1. Imputed rental value/ Rent of ponds	3178.60	14.81	3842.02	17.82	2911.99	17.89	3306.61	17.35
2. Interest on fixed capital	931.53	4.34	581.98	2.70	446.94	2.75	570.45	2.99
3. Depreciation	1786.70	8.32	1130.23	5.24	931.80	5.72	1134.51	5.95
Total Fixed Cost (B)	5896.83	27.47	4554.23	25.76	4290.73	26.36	5011.57	26.29
Total Cost (A+B)	21464.25	100.00	21558.29	100.00	16276.01	100.00	19062.94	100.00

ponds (Rs. 15.57 thousand) and big ponds (Rs. 11.99 thousand). Among the major variable costs incurred in fish production, the comparatively higher expenses was incurred on human labour (Rs. 5.51 thousand), followed by finger lings (Rs. 3.84 thousand) and lime (Rs. 1.28 thousand), which constituted 28.92 per cent, 20.12 per and 6.73 per cent of total cost, respectively (Table 5.1). Manures, fertilizers, toxicants and feed were still not important items of expenditure in fish production in north Bihar. The expenses on packaging materials and marketing jointly constituted 5.35 per cent of total cost incurred in fish production in the project area. Miscellaneous expenses include, religious expenses and RANGDARI fee which worked out to be Rs. 1247.87 per hectare. It was comparatively higher on medium size of ponds (Rs. 1456.00) followed by large ponds (Rs. 1162.99) and small ponds (Rs. 901.51) however the proportionate expenses was comparatively higher on large ponds (7.14%) which declined with the decline in the size of ponds.

While analysing the pond-size wise expenses on different variable cost items, it was found that per hectare expenses on human labour, fingerlings and manures were comparatively higher on small size of ponds which declined with the increase in the size of fish ponds, that is; the higher the size of ponds, the lower per ha expenses on these items. On the other hand, per ha expenses on packaging materials and marketing were comparatively higher on large ponds (Rs. 462.54 and Rs. 801.22 respectively) which declined with the decline in the size of ponds. It was only due to fact that large pond owners preferred to sell fish in district level market and they had to make expenses on packaging and marketing. Per ha expenses on lime, fertilizers and feed were comparatively higher on medium ponds (Rs. 149.33, Rs. 1530.77 and Rs. 200.19, respectively), followed by on big ponds (Rs. 100.87, Rs. 1174.75 and Rs. 182.53 respectively) and small ponds (Rs. 120.61, Rs. 990.36 and Rs. 138.16, respectively). Small ponds owners did not use toxicant in the fish ponds however, medium and big pond operators used toxicants as preventive measures against diseases of fish.

It may be inferred from above discussion that the per hectare cost of fish production was comparatively higher on smaller size of ponds (medium and small) and lower on large ponds, mainly due to use of larger number of human labour and fingerlings on the smaller size of ponds. However, packaging and marketing costs were higher

on large ponds, mainly due to market oriented production on these
ponds.

Per kg cost of fish production was also estimated for all the
three categories of ponds which are presented in Table 5.2.

It may be observed from the table that, on an average, per kg.
cost of fish production was Rs. 13.46, constituting Rs. 3.34 fixed cost
and Rs. 9.92 variable cost. However, the comparatively higher per
kg cost of fish production was on small ponds (Rs. 16.86) followed
by Medium (Rs. 13.47) and large ponds (Rs. 12.41). Per kg fixed and
variable costs incurred on fish production had the similar trend,
that is; declined with the increase in the size of ponds. The
comparatively higher cost of fish production on smaller ponds was
mainly due to higher per kg fixed cost on smaller ponds and use of
comparatively higher number of human labour and fish seeds on
these ponds because small ponds were generally operated by poor
farmers who employed more number of family labour in fish
production and used larger quantity of fish seeds mainly collected
from rivers which were prone to higher mortality. It may be noted
that the majority of small pond owners used locally collected fish
seeds and paid almost equal price prevailing for good quality fish
seeds in the project area.

**Table 5.2: Per kg cost of fish production on different categories of
fish ponds.**

Categories of Pond	Cost of Production (Rs./kg)	Fixed Cost (Rs./kg)	Variable Cost (Rs./kg)
Small Ponds	16.86	4.63	12.23
Medium Ponds	13.47	3.47	10.00
Large Ponds	12.41	3.27	9.14
Average	13.46	3.54	9.92

Costs of fish production per hectare of pond and per kg were
also estimated following cost concept of cost A_1, cost A_2, cost B and
cost C which are presented in Table 5.3.

It may further be observed that the per hectare cost A_1, was
comparatively much higher on medium size ponds (Rs. 14.93),
indicating use of higher quantum of purchased inputs on these

ponds. As mentioned above, per hectare Cost A_1 was higher on medium ponds but per kg Cost A_1 was lowest on medium ponds (Rs. 9.33) than other two categories of ponds under investigation. It was mainly due to higher per hectare pond fish production on medium size of ponds.

Hence it may be inferred, on the basis of per kg cost of fish production data, that the medium size pond owners appear to be more efficient in fish production in north Bihar which may further be ascertained through further detailed analysis.

Table 5.3:   Cost of fish production per hectare/per kg as per different cost concept

Cost Concepts	Small Ponds	Medium Ponds	Large Ponds	Average
Cost of Fish Production Per ha Pond (in thousand Rs.)				
Cost A_1	12.08	14.93	12.66	13.43
Cost A_2	15.26	18.77	15.57	16.74
Cost B	16.19	19.35	16.02	17.31
Cost C	21.46	21.56	16.28	19.06
Cost of Fish Production Per kg (in Rs.)				
Cost A_1	9.49	9.33	9.65	9.48
Cost A_2	11.99	11.72	11.87	11.82
Cost B	12.72	12.09	12.21	12.22
Cost C	16.86	13.47	12.41	13.46

It may be observed from the table that the total cost of fish production was much higher on small fish of ponds (Cost C – Rs. 21.46 thousand) than large ponds (Rs. 16.26 thousand) but the paid out cost (Cost A_1) was comparatively lower on small ponds than large and medium ponds. The difference of cost B and Cost C was comparatively higher on small ponds (Rs. 5.27) than that of medium ponds (Rs. 2.21 thousand) and large ponds (0.26 thousand). It clearly indicates use of a larger number of family labour in fish production on small and medium size of ponds. Hence, it may be said that the farmers owning/operating small fish ponds used very low level of purchased inputs and larger number of human labour in fish production, indicating more use of traditional inputs in fish production on smaller size of ponds.

Use of major inputs

In traditional fish farming, human labour and fish seeds were important inputs used in fish-production. In present situation also, these two inputs constitute major part of total inputs but other inputs like, lime, manures, fertilizers, feeds, and toxicants are being used by a large number of fish farmers. In the present sub-section, the use of two major inputs namely human labour and fish seeds in fish production has been analysed in detail to have an idea about their extent of use in the project area. The use of these two inputs on different size of ponds have been examined and discussed in detail.

Human labour

In the present investigation, human labours are categorised in two groups that is family labour and hired labour. Hired labour includes attached labour, contractual labour and casual labours. As mentioned in methodology, female and child labours were converted in adult Male labour on the basis of wages paid to them. Operation-wise per hectare number of human labours (family and hired) employed in fish production were computed which are presented in Table 5.4.

Table 5.4: Per hectare utilization hired of and family labour in fish production on different categories of ponds

(in days)

Size of Ponds	Family Labour	Hired Labour	Total
Small	186.96	58.87	240.83
	(75.56)	(24.44)	(100.00)
Medium	69.79	113.83	183.62
	(38.01)	(61.99)	(100.00)
Large	8.24	92.26	100.50
	(8.19)	(91.81)	(100.00)
Average	57.53	95.54	153.07
	(37.58)	(62.42)	(100.00)

Figures in parentheses indicate percentage to total labour utilization on respective cateogry of ponds.

It may be observed from the table that the per hectare number of human labour employed in fish production was 153.07 days however it was comparatively higher on small ponds (240.83 days) followed by medium ponds (183.62 days) and large ponds (100.50 days). Hired human labour constituted about 62.42 per cent of total human labours employed in fish production. Number of hired human labour employed in fish production was comparatively higher on medium ponds (113.83 days) but the proportion of hired labour in total human labours employed was comparatively higher on large ponds (91.81 per cent) followed by medium ponds (62.00 per cent) and small ponds (24.44 per cent).

The comparatively higher employment of human labour on small size of ponds was mainly due to higher use of family labour in fish production on this particular size of ponds because most of them (90 per cent) had fish production as main occupation (Table 3.5) and they might not had other opportunities of gainful employment.

Despite the abundance of family labours, hired human labours were employed on all the size of ponds because some of the operations like; stocking and harvesting are very specialised activities which could be done by skilled labours in limited time. Hence, fish farmers of all the size groups had to employ hired human labours for performing specialised activities in fish production.

Operation wise employment of human labour has been computed to examine its extent of utilization in different activities of fish production. The related data are presented in Table 5.5 and 5.6.

It may be observed from the table that the comparatively larger number of human labour was employed in 'Watch and Ward' (73.51 days), followed by harvesting (28.08 days), input application (26.92 days), stocking (11.30 days), marketing (6.53 days), pre-harvesting (4.70 days) and pre-stocking (2.02 days). The trend holds true on all size of ponds under investigation.

The watch and word seems to have no direct relation with the level of fish production but all the pond owners made a comparatively higher expenses on this particular operation which varied from 44.79 per cent of total expenses on human labour on small ponds to 51.14 per cent on large size of ponds (Table 5.1). Moreover, per hectare expenses on 'watch and ward' declined with the increase in the size of the pond. It may further be observed that the 'watch and ward' was the domain of family labour on small ponds but medium and large pond operators had to hire in human labour for this particular operation. Farmers are, no doubt, making

Table 5.5: Operation-wise per hectare human labour utilization on different categories of ponds

(in man days)

Operation	Small			Medium			Large			Average		
	FL	HL	TL	FL	HL	TL	FL	HL	TL	FL	HL	TL
Pre-stocking	1.04 (77.61)	0.30 (22.39)	1.34 (100.00)	1.12 (44.09)	1.42 (55.91)	2.54 (100.00)	0.20 (11.04)	1.61 (88.96)	1.81 (100.00)	0.68 (33.66)	1.34 (66.34)	2.02 (100.00)
Stocking	14.73 (64.49)	8.11 (35.51)	22.83 (100.00)	3.02 (21.89)	11.23 (78.81)	14.25 (100.00)	1.11 (21.14)	4.14 (78.86)	5.25 (100.00)	3.86 (34.15)	7.44 (65.85)	11.30 (100.00)
Input Application	40.93 (85.54)	6.92 (14.46)	47.85 (100.00)	6.75 (21.42)	24.75 (78.58)	31.50 (100.00)	1.36 (8.19)	15.24 (91.81)	16.60 (100.00)	9.29 (34.50)	17.63 (65.50)	26.92 (100.00)
Watch & Ward	104.55 (93.54)	7.22 (6.46)	111.77 (100.00)	45.56 (52.37)	41.42 (47.63)	86.98 (100.00)	2.63 (5.20)	47.85 (94.8)	50.48 (100.00)	34.15 (46.45)	39.36 (53.55)	73.51 (100.00)
Pre-harvesting	1.09 (31.96)	2.32 (68.04)	3.41 (100.00)	2.08 (40.94)	3.00 (59.06)	5.08 (100.00)	0.37 (7.72)	4.42 (92.28)	4.79 (100.00)	1.13 (24.04)	3.57 (75.96)	4.70 (100.00)
Harvesting	11.81 (26.20)	33.27 (73.80)	45.08 (100.00)	6.06 (16.92)	29.75 (83.08)	35.81 (100.00)	0.81 (4.93)	15.61 (95.07)	16.42 (100.00)	4.45 (15.85)	23.63 (84.15)	28.08 (100.00)
Marketing	7.81 (91.35)	0.74 (8.65)	8.55 (100.00)	5.21 (69.83)	2.25 (30.17)	7.46 (100.00)	1.75 (34.00)	3.39 (66.00)	5.14 (100.00)	3.97 (60.79)	2.56 (39.21)	6.53 (100.00)
Overall	181.96 (75.56)	58.87 (24.44)	240.83 (100.00)	69.79 (38.00)	113.83 (62.00)	183.62 (100.00)	8.24 (8.19)	92.26 (9181)	100.50 (100.00)	57.53 (37.58)	95.54 (62.42)	153.07 (100.00)

Figures in parentheses indicate percentage to total labour utilization on respective category of ponds.

Table 5.6: Operation-wise per hectare cost incurred on human labour on different categories of ponds

(in man days)

Operation	Small			Medium			Large			Average		
	FL	HL	TL	FL	HL	TL	FL	HL	TL	FL	HL	TL
Pre-stocking	30.10 (0.57)	9.94 (0.51)	40.04 (0.55)	33.46 (1.52)	45.23 (1.20)	78.69 (1.32)	6.30 (2.46)	51.71 (1.64)	58.01 (1.70)	20.20 (1.16)	43.04 (1.34)	63.24 (1.28)
Stocking	405.09 (7.68)	239.25 (12.19)	644.34 (8.90)	82.63 (3.74)	321.85 (8.54)	404.48 (6.77)	31.07 (12.13)	31.07 (3.86)	121.47 (4.48)	152.54 (6.09)	106.30 (6.71)	215.45 (6.49)
Input Application	1165.79 (22.11)	221.85 (11.31)	1387.64 (19.18)	193.13 (8.75)	751.87 (19.95)	945.00 (15.81)	37.66 (14.71)	476.82 (15.14)	514.48 (15.11)	264.55 (15.15)	543.95 (16.95)	808.50 (16.31)
Watch & Ward	3001.48 (56.92)	239.69 (12.22)	3241.17 (44.79)	1459.02 (66.08)	1411.35 (37.46)	2870.37 (48.03)	86.64 (33.83)	1654.81 (52.55)	1741.45 (51.14)	1043.41 (59.75)	1351.71 (42.12)	2395.12 (48.33)
Pre-harvesting	34.65 (0.66)	81.31 (4.14)	115.96 (1.60)	69.60 (3.16)	108.00 (2.87)	177.69 (2.97)	12.20 (4.76)	160.51 (5.10)	172.71 (5.07)	37.48 (2.15)	128.70 (4.01)	166.18 (3.35)
Harvesting	388.19 (7.36)	1144.59 (58.33)	1532.78 (21.18)	202.12 (9.15)	1051.15 (27.90)	1253.27 (20.97)	25.32 (9.89)	565.96 (17.97)	591.28 (17.36)	146.70 (8.40)	837.11 (26.09)	983.81 (19.85)
Marketing	248.20 (4.71)	25.46 (1.30)	273.66 (3.78)	167.81 (7.60)	78.42 (2.08)	246.23 (4.12)	56.90 (22.22)	117.91 (3.74)	174.81 (5.13)	127.65 (7.31)	89.10 (2.78)	216.75 (4.37)
Overall	5273.50 (100.00)	1962.09 (100.00)	7235.29 (100.00)	2207.86 (100.00)	3767.87 (100.00)	5975.73 (100.00)	256.09 (100.00)	3149.20 (100.00)	3405.29 (100.00)	1746.29 (100.00)	3209.06 (100.00)	4955.35 (100.00)

Figures in parentheses indicate percentage to total labour utilization on respective category of ponds.

heavy expenses on 'watch and ward' in fish production but it can not be avoided due to poor law and order situation in Bihar.

Fish harvesting was the second most important operation with respect to utilization of human labour. About 18.34 per cent of total labour employed in fish production was engaged in harvesting operation. However, per hectare use of human labour declined with the increase in the size of ponds. This operation was conducted mainly by hired labour because hired labour constituted 84.15 per cent of total labour engaged in this particular operation. It may be pointed out that large pond owners performed the harvesting operation through hired labour exclusively because per hectare family labour employed in fish harvesting was less than one day.

Per hectare expenses on human labour for fish harvesting was worked out to be Rs. 983.81, accounting for about 19.85 per cent of total cost of labour incurred in fish production however the amount of expenditure on human labour and its proportion to total expenditure on human labour declined with the decline in the size of ponds in the project area.

Input application emerged as third important operation in fish production with respect to human labour employment. Besides, application of various inputs in ponds human labours were also required for transportation of manures, lime, fertilizers, feed and toxicants from market to farmhouse/ponds. Per hectare human labour employed in this operation was worked out to be 26.92 man days, constituting 9.29 family labours and 17.63 hired labours which accounted for 34.50 per cent and 65.50 per cent, respectively of total labour employed in this operation. It has further been observed that per hectare human labour employed in input application declined with the decline in the size of ponds.

Per hectare cost incurred on human labour in this operation was worked to be Rs. 808.50, constituting 16.31 per cent of total expenditure made in fish production however the proportionate expenses on input application has been approximately 15 per cent on medium and large size ponds and about 19 per cent on small size ponds.

As discussed earlier, the stocking is most important operation in fish-culture. Stocking means putting fish seeds to the pond. The stocking activity starts from bringing fish seeds from traders, Govt. and private hatcheries to putting fish seeds to pond. While bringing fish seeds from different sources to the pond, family or hired labours are required to take precaution for less mortality which could be achieved by making arrangement for good aeration. Before putting fish seeds to the pond some of the fish farmers used solution of $KMNO_4$ for dipping the fish seeds before putting them to the pond. On the sample ponds under investigation, about 11.30 human labour (man days) were employed in stocking of fish seeds in ponds, comprising 3.86 family labour days and 7.44 hired labour days which accounted for 34.15 per cent and 65.85 per cent of total human labour employed in this particular operation. This operation was done through mainly hired labour (78.80%) on medium and large ponds, and small pond owners had also employed about 8 hired human labours for this particular operation. It may be pointed out that the stocking operation can only be done by skilled labour. Hence, hired human labour is required on all size of ponds for this operation. Fish farmers, on an average, incurred expenses of about Rs. 258.84 per hectare on human labour for stocking operation which increased with the decline in the size of ponds.

Fish production is a commercial venture because it is produced for market. Hence, the marketing has now emerged as an important activity in fish production. Per hectare, 6.53 human labour man days were employed in marketing of fish in the project area which constituted 60.79 per cent family human labour and 39.21 per cent hired human labour. It is only operation which was performed mainly by family labours on small ponds (91.35%) and medium pond (69.83 per cent). The large pond owners also performed this operation by one third of family labours. Hence, it may be said that the marketing operation is the domain of family labour in the project area. It may be due to fact that the marketing operation does not need any skill and it is not strenuous job hence, it was preferred by farmers to perform the job by their family members. Moreover, it involves transaction of money and fish farmers performed this job by themselves to avoid any financial lapses by hired labours.

Pre stoking and pre-harvesting are minor operations and per hectare about 7 human labour days, were required to performance these two operations.

On the basis of above discussions it may be inferred that the fish farming is labour intensive enterprise. About 50 per cent of human labour is required to protect the fish from theft, mainly due to poor law and order situation of the state. It may further be concluded that the small pond owners managed their fish production activities through family labours whereas large pond operators had to depend on hired labours for fish production. It was mainly due to large size of ponds which could not be managed by family labours. Moreover, some of the large ponds owners are rich and they do not like to work in pond's water which is comparatively strenuous jobs.

Fish seeds

The species-mix in fish seeds and stocking rate are two important determinants in economics of pisi-culture that is; cost of production and level of output. In the project area, composite carp culture has been commonly practised which includes major carps, exotic carps, minor carps and cat fish. In the present section, an effort has been made to examine the species mix and stocking rate in the project area.

Fish seeds stocked in the sample ponds have been categorised in two groups that is; indigenous carps and exotic carps. The former includes major carps, minor carp and cat fish whereas later includes exotic carps that is; common carp, Grass Carp and silver carp. Per hectare quantity and number of fish seeds stocked on three categories of ponds were computed which are presented in Table 5.7.

Table 5.7: Per hectare quality and number of Fish seeds stocked on different categories of ponds

(Quantity in kg and number in 000')

Species	Small		Medium		Large		All ponds	
	Qty.	Number	Qty.	Number	Qty.	Number	Qty.	Number
Indigenous	20.27	19.12	16.58	13.86	12.85	10.54	15.37	13.08
	(76.63)	0.943*	(75.88)	0.836	(79.66)	0.820	(77.47)	0.851
Exotic	6.18	4.99	5.27	4.07	3.28	2.56	4.47	3.50
	(23.37)	0.807	(24.12)	0.772	(20.34)	0.780	(s 2.53)	0.783
Total	26.45	24.11	21.85	17.93	16.13	13.09	19.84	16.58
	(100.00)	0.912	(100.00)	0.821	(100.00)	0.812	(100.00)	0.836

(i) Figures in parentheses indicate percentage to total quantity of fingerlings in respective category of ponds.
*   Figures indicate per kg number of fingerlings.

It may be observed from the table that on an average, per hectare 16.58 thousand of fish seeds, amounting 19.84 kg were stocked in the project area which constituted 77.47 per cent indigenous and

22.53 per cent exotic fish species. Per hectare rate of stocking was comparatively higher on small size of ponds (26.45 kg) followed by medium ponds (21.85 kg) and large ponds (16.13 kg). The higher quantum of fish seeds by small pond owners for stocking was mainly due to use of poor quality of fish seeds by them which were collected from natural sources which had higher mortality. Small pond owners had low liquidity and were not in position to purchase quality fish seeds. It has further been observed that the large size pond owners used comparatively larger size of fish seeds (820 fry per kg of fish seeds) than medium pond owners (836 fry per kg of fish seeds) and small ponds owners (943 try per kg fish seeds). But the medium ponds owners stocked fish ponds by comparatively larger proportion of exotic species of fish seeds (24.12 per cent) than the corresponding proportion on small ponds (23.37 per cent) and large ponds (20-34 per cent).

To have a deep understanding of composition of different species of indigenous carps, pond size-wise stocking of common indigenous fish species namely; Rohu, Catla, Mrigal and others (Mangur, Boari, Garai etc.) were computed which are presented in Table 5.8.

Table 5.8: **Per hectare quantity and number of Indigenous fingerlings on different categories of ponds**

(Quantity in kg and number in 000')

Species	Small		Medium		Large		All ponds	
	Qty.	Number	Qty.	Number	Qty.	Number	Qty.	Number
Rohu	7.46	6.96	5.92	4.94	4.63	3.79	5.54	4.70
	(36.80)	0.932*	(35.71)	0.834	(36.03)	0.819	(36.04)	0.848
Catla	6.88	6.48	4.26	3.56	3.48	2.89	4.03	3.47
	(33.94)	0.941	(33.17)	0.835	(33.00)	0.822	(33.25)	0.852
Mrigal	5.19	5.05	4.26	3.56	3.48	2.89	4.03	3.47
	(25.60)	0.974	(25.69)	0.837	(27.08)	0.829	(26.22)	0.860
Others	0.74	0.64	0.90	0.76	0.50	0.377	0.69	0.56
	(3.66)	0.870	(5.43)	0.848	(3.89)	0.755	(4.49)	0.818
Total	20.27	19.12	16.58	13.86	12.85	10.54	15.37	13.08
	(100.00)	0.943	(100.00)	0.836	(100.00)	0.820	(100.00)	0.851

* Figures in parentheses indicate percentage to total quantity of fingerlings in respective category of ponds.
* Figures indicate per kg number of fingerlings.

It may be observed from the table that the comparatively larger proportion of Rohu (36.04 per cent) was stocked in the project area followed by Catla (26.22 percent), Mrigal (26.22 per cent) and others (4.49 per cent). The proportions of Rohu, Catla and Mrigal were almost same on all size of ponds but the comparatively larger proportion of Maugur (Major species in other category) was stocked in medium size ponds. It was mainly due to increased interest amongst medium pond owners about production of Mangur fish since it fetches higher price in the market. Rohu was also preferred due to comparatively higher demand and greater conversion ratio than that of catla and marigal.

Table 5.9: **Per hectare quantity and number of Exotic fingerlings on different categories of ponds**

(Quantity in kg and number in 000')

Species	Small		Medium		Large		All ponds	
	Qty.	Number	Qty.	Number	Qty.	Number	Qty.	Number
Common	2.47	2.22	2.40	1.89	1.14	0.95	1.82	1.50
Carp	(39.97)*	0.897*	(45.54)	0.787	(34.76)	0.833	(40.72)	0.822
Silver Carp	1.68	1.35	1.04	0.86	0.75	0.62	1.00	0.82
	(27.18)	0.804	(19.73)	0.828	(22.87)	0.828	(22.37)	0.821
Grass Carp	2.03	1.42	1.83	1.32	1.39	0.99	1.65	1.18
	(32.85)	0.702	(34.73)	0.722	(42.37)	0.710	(36.91)	0.715
Total	6.18	4.99	5.27	4.07	3.28	2.56	4.47	3.50
	(100.00)	0.807	(100.00)	0.772	(100.00)	0.780	(100.00)	0.783

*   Figures in parentheses indicate percentage to total quality of fingerlings in respective category of ponds.
**   Figures indicate per kg number of fingerlings

Among exotic species of fish, the common carp, on an average, constituted comparatively larger proportion (40.72 per cent) followed by grass carp (36.91 per cent) and silver carp (22.37 per cent) in the project area (Table 5.9). Per hectare stocking rate of common carp varied from 2.47 kg on small ponds to 2.40 kg on medium ponds and 1.14 kg on large ponds. Per hectare stocking rate of common carp was, nodoubt, comparatively higher on small ponds, mainly due to their higher stocking rate but its proportion to total exotic species of fish seeds stocked in ponds under investigation was higher on medium ponds. Hence, it may be said that the common carp was most important exotic species of fish on small and medium size of ponds. Grass carp was the second important species of fish seeds for stocking in ponds under investigation. Per hectare stocking rate of this particular species of fish was comparatively higher on small ponds (2.03 kg) followed by medium ponds (1.83 kg) and large ponds (1.39 kg) but the proportion grass carp fish seeds to total exotic species of seeds had just reverse trend, that is the comparatively higher proportion on large ponds (42.37 per cent) followed by medium ponds (34.73 per cent) and small ponds (32.85 per cent).

An effort has been made to examine the proportions of different species of fish *vis–a-vis* their recommended proportions for stocking in multi-carp culture. All the six major carps that is, rohu, catla, mrigal, common, silver and grass carps have been categorised in four major groups. These are upper level carps (catla + silver) middle level (rohu), lower level (common + mrigal) and grass eating carp (grass carp). The stocking of indigenous carp like minor carps and cat fishes was not common hence the proportion of these fish were calculated but not discussed in detail.

The proportions of four major groups of fish stocked in the three specified categories of ponds were computed which are presented in Table 5.10.

Table 5.10: Proportions of groups of fish stocked in different categories of ponds under investigation

(Percentage)

Groups of Fishes	Recommended proportion	Small	Medium	Large	Average
Surface feeder (Catla + Silver Carp)	30 - 40	32.33	29.93	30.91	30.78
Column feeder (Rohu)	10 - 20	28.22	27.11	28.68	27.93
Bottom feeder (Common Carp + Mrigal)	40 - 45	28.98	30.45	28.68	29.49
Grass eating fishes (Grass Carp)	5 - 15	7.66	8.36	8.62	8.32
Others	-	2.81	4.14	3.10	3.48

It may be observed from the table that the fish farmers stocked about 28 per cent of rohu- the middle level fish species against the recommended proportion of 10-20 per cent. The proportions of surface feeder fish (catla + silver carp) and grass eating fish (grass carp) were 30.78 per cent and 8.32 per cent, respectively which almost matched the recommended proportion of 30-40 per cent and 5–15 per cent, respectively. The proportion of bottom feeder fish (common carp + marigal) was worked out to be 29.49 per cent against the recommended proportion of 40-45 per cent. Different size of ponds did not differ much with respect to proportion offish of different levels.

On the basis of above discussions it may be inferred that the stocking rate was much higher in the project area. It was mainly due to stocking of small size of fish seeds of high mortality rate because there is a dearth of fish nurseries in Bihar. There are only five Government and four private fish nurseries which produced fish seeds of232 million in the year 1995-96. These fish are not sufficient for stocking in one-third of ponds area under fish culture. While conducting survey we got an opportunity to discuss with private fish growers but they were unaware of scientific method of fish seed production. Moreover, fish farmers of the project area do not prefer to purchase fish seed from them due to poor quality, mix species and higher prices charged by them.

Sources of fish seeds

In this section, sources of fish seeds of sample fish farmers have been examined to have an idea about the importance of sources of fish seeds in the project area. There were three major sources of fish seeds, namely; fish farmers, trader and fish co-operative. Number of farmers procured fish seeds from these sources were computed which are presented in Table 5.11.

Table 5.11: Sources of fish seeds of the farmers

(Number of farmers)

Source	Indigenous Major Carps	Mangur	Exotic Carps
Fish farmers	17	4	7
Traders	142	22	139
Fisheries cooperative societies	21	-	12
Total	180	26	158

It may be observed from the table that the farmers utilised multi sources for procuring fish seeds but the fish seeds traders emerged as the most important source since 142 fish farmers (78.89%) procured indegeneous major carps, 22 fish farmers (12.22%) procured mangur and 139 fish farmers (77.22%) procured exotic carps from them. Fish co-operative societies are dead organization in north Bihar but 21 fish farmers procured exotic carps from co-operatives.

Fish farmer was not the major source of fish seeds in the project area. There were only 4 fish farmers who collected fish seeds from rivers (specially the river Ganges) and sold to other fellow farmers. However, 17 fish farmers provided fish seeds to fellow farmers but they were not involved in trading of fish seeds. They purchased fish seeds more than their requirements and surplus fish seeds were sold by them to fellow farmers. Hence, it may be said that the trader was the main source of fish seeds in the project area.

While analysing the purchase place of fish seeds, about 50 per cent of farmers could get indigenous and exotic carps on ponds since traders used to deliver fish seeds on fish ponds (Table 5.12). Local market was most important source for mangur fish seeds because 42.31 per cent of Mangur growing farmers purchased seeds in the local market. However it was second important source for indigenous and exotic carps in the project area. Farmers also purchased fish seeds from outside state (West Bengal) but their proportions were less than 10 per cent for indigenous and exotic carps but about 31 per cent maugur growing farmers purchased fish seeds from out side state.

Table 5.12: Place of Procurement of fish seeds

(Number)

Place	Farmers who procured		
	Indigenous carps	Mangur	Exotic carps
Ponds delivery	98 (54.44)	3 (11.54)	82 (51.90)
Local village	27 (15.00)	4 (15.38)	19 (12.02)
Local Market	38 (21.11)	11 (42.31)	44 (27.85)
Outside the State	17 (9.44)	8 (30.77)	13 (8.23)
Total	180 (100.00)	26 (100.00)	158 (100.00)

While conducting survey it was observed that out of 180 fish farmers under study 59 farmers changed the source of fish seeds. Reasons for change in fish seed sources are poor quality, higher prices, distant place, inadequate availability, delayed in availability and transportation problems. Among reasons for change in source of fish seeds, poor quality was the most important reason followed by delayed availability higher price, inadequate availability, transportation problem and distance places (Table 5.13).

Table 5.13: Ranks of reasons for changing the source of fish seeds

	Reasons	Rank
1.	Poor quality	1.95
2.	Delayed availability	2.21
3.	Higher price	3.15
4.	Inadequate availability of desired species of seeds	3.43
5.	Transportation problem	3.67
6.	Distant place	3.81

It has further been observed that the farmers considered poor quality of fish seeds due to improper size, higher mortality rate and mixed species. Among the above three determinants of poor quality, the high mortality rate was the most important determinant of poor quality of seeds followed by mixed species and small size of fish seeds (Table 5.14).

Table 5.14: Reasons for poor quality of fish seeds

	Reasons	Rank
1.	Higher mortality	1.87
2.	Mixed species	2.13
3.	Small size	2.86

On the basis of above discussions it may be surmised that the per hectare stocking rate was much higher in the project area. The composition of different species of fish seeds in multi-carp culture was mismatched to the recommendation. The rohu was most preferred species and the stocking rate of this particular species of fish was higher than requirement.

The reasons for high stocking rate may be traced from the use of small size of fish seeds and their high mortality rate (Table 5.14). Finally, it may be said that the fish farmers are unaware of Scientific method of stocking, particularly the rate, size of fish seeds and composition in multi carp cultivation.

6

FISH PRODUCTION, POTENTIALITY, EFFICIENCY AND PROFITABILITY

Fish Production

The reliable estimate of fish production and productivity are not available for Bihar however per hectare fish productivity was estimated to 2170 kg in ponds which were managed and operated under Fish Farmers Development Agency Project (Govt. of India, 1997).

In the present section, an effort has been made to examine the fish production in the project area. The related data obtained from respondent fish farmers were computed and size group wise production/productivity of indigenous and exotic fish are presented in Table 6.1.

Table 6.1: Per hectare production of fish on different categories of ponds

(Quantity in Qt.)

Species	Small Qty.	Medium Qty.	Large Qty.	All Ponds Qty.
Indigenous	9.24 (72.58)	11.99 (74.89)	9.63 (73.40)	10.47 (73.94)
Exotic	3.49 (27.42)	4.02 (25.11)	3.49 (26.60)	3.69 (26.06)
Total	12.73 (100.00)	16.01 (100.00)	13.12 (100.00)	14.16 (100.00)

Figures in parentheses indicate percentage to per ha total quantity of fish produced in respective category of ponds.

It may be observed from the table that per hectare fish production was worked out to be 14.16 quintals, constituting 73.94 per cent indigenous and 26.06 per cent exotic fish in the project area. The comparatively higher per hectare fish productivity was observed in Medium ponds (16.01 quintals) followed by large ponds (13.12 quintals) and small ponds (12.73 quintals). It may further be observed that the per hectare production of exotic species of fish was identical on small and large ponds (3.49 quintals) but large pond owners could produce 39 kg more indigenous fish than that of small pond owners on per hectare basis. Per hectare productivity of fish was comparatively higher on medium ponds, mainly due to use of comparatively higher fertilizers, lime and supplementary feeds (Appendix-V).

Per hectare production of three major carps and other indigenous carps were also computed which are presented in Table 6.2.

Table 6.2: Per hectare quantity of fish of different indigenous species produced on different categories of ponds

(Quantity in quintal)

Species	Small Qty.	Medium Qty.	Large Qty.	All Ponds Qty.
Rohu	3.29 (35.61)	4.34 (36.20)	3.46 (35.93)	3.77 (36.01)
Catla	2.88 (31.17)	3.96 (33.03)	3.10 (32.19)	3.39 (32.38)
Mrigal	1.94 (21.00)	2.57 (21.43)	2.40 (24.92)	2.40 (22.92)
Others	1.13 (12.23)	1.12 (9.34)	0.67 (6.96)	0.91 (8.69)
Total	9.24 (100.00)	11.99 (100.00)	9.63 (100.00)	10.47 (100.00)

Figures in parentheses indicate percentage to total quantity of indigenous fish species produced in respective category of ponds.

Per hectare production of Rohu was worked out to be 3.77 quintals, constituting 36.01 per cent of production of indigenous carps in the project area. Catla was the second important indigenous fish produced in the project area since production of this particular species was only 38 kg less them per hectare production of Rohu. Among the major carps, per hectare production of marigal was the lowest (2.40 quintals) because it might be the least preferred fish species and low price fetching major carp.

Other category of fish includes several species namely minor carps, mangur and cat fish, Mangur was predominantly grown in Champaran district whereas minor carps and cat fish were grown in other 5 districts under study. Per hectare production of other category of fish was worked out to be 910 kg however it was comparatively higher on small fish ponds and lower on larger ponds, that is 1.12 quintal on medium ponds and 0.67 quintal on big ponds.

Among exotic species of fish, common carp and grass carp jointly constituted about 74 per cent and silver carp constituted about 26 per cent of total exotic fish production in the project area (Table 6.3). Pond size group wise analysis revealed that the grass carp species of fish was produced in larger proportion on big size of fish ponds (40.40%). However, common carp was more common on small and large size ponds. Per hectare production of silver carp was comparatively higher of small ponds (1.03 quintal) than medium (0.96 quintal) and large ponds (0.91 quintal).

Table 6.3: Per hectare quantity of exotic species of fish on different categories of ponds

(Quantity in quintal)

Species	Small Qty.	Medium Qty.	Large Qty.	All Ponds Qty.
Common carp	1.41 (40.40)	1.67 (41.54)	1.17 (33.52)	1.40 (37.94)
Silver Carp	1.03 (29.51)	0.96 (23.88)	0.91 (26.08)	0.94 (25.47)
Grass Carp	1.05 (30.09)	1.39 (34.58)	1.41 (40.40)	1.35 (36.59)
Total	3.49 (100.00)	4.02 (100.00)	3.49 (100.00)	3.69 (100.00)

Figures in parentheses indicate percentage to total quantity of exotic fish produced in respective category of ponds.

On the basis of above discussions, it may be said that the Rohu and catla of indigenous category fish and common carp and grass carp of exotic category of fish were most common species of fish produced in the project area. However, indigenous species of fish is still more common in north Bihar but all the categories of farmers have started production of exotic species of fish in north Bihar.

Conversion Ratio

The production of fish is a function of various variables like fish seeds, human labour, inputs, and management however fish seeds is an indispensable and necessary factor of production. The use of appropriate number and quantity of quality seeds may help obtaining higher level of fish production. In this context conversion ratio is an important measure of production efficiency in fish production because all species of fish do not have identical potential of growth. Hence, the use of higher potential species of fish may yield higher quantity of fish in the project area in general and in a particular pond, in particular.

Keeping in view this objective, an attempt has been made to examine the conversion ratios of different species of fish on sample ponds. Pond size group wise conversion ratios of all major fish species have been worked out and presented in Table 6.4.

It may be observed from the table that, on an average, the conversion ratio was 71.37, indicating that the use of one kg of fish seeds would produce about 71 kg of matured fish. The conversion ratio was considerably higher on big ponds (81.34) which declined with the decline in the size of ponds that is, 73.27 on medium ponds and 48.13 on small ponds. Among different fish species, the comparatively higher conversion ratio was observed in case of 'other categories of indigenous fish (132) than that of all the indigenous and exotic species of fish produced on sample ponds. It does not mean that the indigenous species of fish had higher conversion ratio in true sense because it was mainly due to left over fish seeds of minor indigenous species of fishes in the ponds. Moreover, flood water also brings some fish of minor carps and cat fish into ponds which might have also increased the fish production in ponds resulting in higher conversion ratio.

It has also been observed that, on an average, exotic fish species had comparatively higher conversion ratio (83) than that of indigenous fish species (68) on fish ponds under investigation. Among the exotic fish, Silver carp had the comparatively higher conversion ratio (94) than grass carp (82) and common carp (77) whereas among indigenous major carps, the conversion ratio of Rohu (68) was comparatively higher than Catla (66) and marigal (60). Pond size group wise analysis of conversion ratio of fish species under investigation revealed that the large ponds had much higher conversion ratios for all the species under study which declined

Table 6.4: Fish species-wise conversion ratios on different categories of fish ponds

(Production and fingerlings in kg)

Species	Small ponds			Medium ponds			Large ponds			Average		
	Production	Fingerlings	Conversion ratio	Production	Fingerlings	Conversion ratio	Production	Fingerlings	Conversion ratio	Production	Fingerlings	Conversion ratio
Rohu	329	7.46	44.10	434	5.92	73.31	346	4.63	74.73	377	5.54	68.05
Catla	288	6.88	41.86	396	5.50	72.0	310	4.24	73.11	339	5.11	66.34
Mrigal	194	5.19	38.04	257	4.26	60.33	240	3.48	.68.97	240	4.03	59.55
Other	113	0.74	152.70	112	0.90	124.44	67	0.50	134.00	91	0.69	131.88
Indigenous fish	924	20.27	45.58	1199	16.50	72.32	963	12.85	74.94	1047	15.37	68.12
Common Carp	141	2.47	57.08	167	2.40	69.58	117	1.14	102.63	140	1.82	76.92
Silver Carp	103	1.68	61.31	96	1.04	92.31	91	0.75	121.33	94	1.00	94.00
Grass Carp	105	2.03	51.72	139	1.83	75.96	141	1.39	101.44	135	1.65	81.82
Total exotic fish	349	6.18	56.47	402	5.27	76.29	349	3.28	106.40	369	4.47	82.55
All type of fish	1273	26.45	48.13	1601	21.85	73.27	1312	16.13	81.34	1416	19.84	71.37

with the decline in pond size. However, the pattern of conversion ratio of different fish species on three specified size of ponds under investigation had the similar trend. In other words, it may be said that the pattern of conversion ratio of different species of fish was similar on all categories of ponds, that is the higher conversion ratios for exotic fish species than indigenous species on size of ponds. The conversion ratio of silver carp was of exotic species higher on all categories of fish ponds and it was also true for Rohu of indigenous carp on all categories of ponds.

The conversion ratio is likely to influence the choice of fish species for stocking. The fish farmers are expected to use larger proportion of fish species which has higher conversion ratio. In the project area, the larger proportion of Rohu in stocking was done, probably due to its higher conversion ratio than other indigenous carps. But it is not true for exotic species of fish because silver carp had the comparatively higher conversion ratio but per hectare rate of stocking was the lowest (94 kg/ha). It may be pointed out that the farmers of the project area do not prefer silver carp for fish production because the demand for silver carp is comparatively low, mainly due to poor taste of this fish species. Hence, it may be said that the conversion ratio is not the sufficient factor which influences the selection of fish species for fish production.

Potential Production

The average per ha fish production was estimated to be 14.16 quintals which seems to be much lower than the potential average fish production in north Bihar, particularly ponds, operating under Fish Farmers Development Agencies (2170 kg/ha)*. In the present study also per hectare fish productivity varied widely from less than 10 quintals to more than 20 quintals. To capture the variations in fish productivity, ponds were categorised on the basis of fish productivity level and their quantum of fish production was computed for all categories. Fish ponds were categorised in four groups on the basis of per hectare fish productivity, that is; less than 10 quintals, 10 – 15 quintals, 15-20 quintals and 20 quintals and above. Number of ponds in all categories of productivity levels alongwith their area in the and production in quintals were computed which are presented in Table 6.5.

---

* Handbook on Fisheries Statistics, Ministry of Agriculture, Fisheries Division, Government of India, 1996, p. 148.

Table 6.5: Distribution of ponds, total water area and total production according to productivity range on different categories of ponds

(Area in ha and production in quintal)

Productivity Range (Qt/ha)	Small ponds			Medium ponds			Large ponds			All Ponds		
	No. of Ponds	Area	Production	No. of Ponds	Area	Production	No. of Ponds	Area	Production	No. of Ponds	Area	Production
< 10	19 (25.0)	5.90 (29.16)	43.60 (16.92)	9 (11.39)	5.25 (10.10)	46.33 (5.56)	6 (12.77)	11.18 (17.47)	99.86 (11.90)	34 (16.83)	22.33 (16.39)	189.79 (9.84)
10-15	33 (43.42)	8.40 (41.52)	102.44 (39.76)	31 (39.24)	20.64 (39.69)	252.86 (30.37)	27 (57.45)	35.80 (55.95)	435.25 (51.82)	91 (45.05)	64.84 (47.61)	790.55 (40.97)
15-20	16 (21.05)	4.02 (19.87)	69.82 (27.10)	19 (24.05)	11.79 (22.67)	206.21 (24.77)	14 (29.79)	17.01 (26.58)	304.29 (36.25)	49 (24.26)	32.82 (24.09)	580.32 (30.08)
20 & Above	8 (10.53)	1.91 (9.44)	41.78 (16.22)	20 (25.32)	14.323 (27.54)	327.13 (39.29)	-	-	-	28 (13.86)	16.23 (11.91)	368.91 (19.12)
Total	76 (100.00)	20.23 (100.00)	257.64 (100.00)	79 (100.00)	52.00 (100.00)	832.53 (100.00)	47 (100.op)	63.99 (100.00)	839.40 (100.00)	202 (100.00)	136.22 (100.00)	1929.57 (100.00)

It may be observed from the table that about 16.83 per cent of ponds had 16.39 per cent of total pond's area and fish productivity of less than 10 quintals per hectare but produced only 9.84 per cent of total fish production on study ponds whereas 13.86 per cent of ponds with 11.91 per cent of ponds area had per hectare fish productivity of 20 quintals and produced about 19.12 per cent of total fish production on ponds under study. However, about 70 per cent of ponds area had per hectare fish productivity of 10–20 quintals and produced almost identical proportion of fish (71 per cent) in the project area. The above information clearly indicate skewedness in fish production. The fish productivity varied more widely on small ponds where 29.16 per cent of small ponds area produced only 16.92 per cent of fish and 9.44 per cent area of small ponds produced about 16.22 per cent of fish produced on this particular size of ponds. None of the large ponds had per hectare fish productivity of 20 quintal and more. Per hectare productivity on 55.45 per cent area of large ponds was 10–15 quintals and these ponds produced 51.82 per cent of fish produced on this category of ponds under study. However, 26.58 per cent area of large ponds had per hectare productivity of 15 – 20 quintals and produced about 36.25 per cent of fish produced on this particular category of ponds. The comparatively larger area of medium category of ponds (27.54%) had per hectare fish productivity of 20 quintals and more and produced 39.29 per cent of fish produced on medium category of fish ponds. It is worth pointing out that only 10.10 per cent of area of medium ponds had fish productivity of less than 10 quintals and produced only 5.56 per cent of fish produced on this category of ponds.

On the basis of above analysis it may be inferred that the larger quantum of fish produced on medium ponds were from the high productive ponds whereas the larger quantum of fish produced on small and large ponds were from low productive ponds of these categories. Per hectare fish productivity of 20 quintals and more on 25.32 per cent of medium ponds and 10.53 per cent small ponds clearly indicates potentiality of fish production in the project area. Some of small and medium ponds holders demonstrated the potentiality in fish production in low fish productive area of north Bihar. Hence, there is a need to have concerted efforts in making the resources and technology available to fish farmers for increasing fish productivity in the State.

Production Efficiency

The fish production is a function of various factor inputs. More specifically, the fish production function refers to the relationship between fish output and input factors namely; fish seeds, human labour, fertilizers, manures, supplementary feeds, lime, toxicants, management etc. In the present section, an attempt has been made to estimate and examine the production elasticities of various factors. We formulated the following null hypothesis "the various input factors (independent variables) included in the analysis do not affect the fish production in the project area". The cobb-Douglas function is used for estimating production elasticities. The following model has been used.

$$Y = ax_1^{b1} \cdot x_2^{b2} \cdot x_3^{b3} \cdot x_4^{b4} \cdot x_5^{b5} \cdot x_6^{b6} \cdot e^u$$

where,

a	=	Intercept
X_1	=	Manures (Rs./ha)
X_2	=	Fertilizers (Rs./ha)
X_3	=	Lime (Rs./ha)
X_4	=	Fingerlings (Rs./ha)
X_5	=	Feed (Rs./ha)
X_6	=	Human labour (wages in Rs./ha)
e^u	=	Error term
$b_1 - b_6$	=	regression co-efficient of respective variables

Analysis is based on cross sectional pond wise per hectare value of output (dependent variable 'Y') and values of all the six important independent variables mentioned above. We have tried liner function and power function by taking dependent and independent variables in physical term. None of models/procedures found appropriate for given data because production of fish did not increase in linear term due to increase in use of inputs. The Cobb-Douglas production function was found comparatively more appropriate which was judged on the basis of F-value and intercept.

The calculated F-value, intercept, multiple-R and coefficient of determination, are presented in Table- 6.6.

The calculated F-value is 13.50 which is greater than the table value of F (2.802) and significant at one per cent level of probability.

It clearly suggests the rejection of null hypothesis, meaning thereby that the independents variables included in the study affect fish production significantly.

Table 6.6: Calculated F-value, co-efficient of determination (R-square), multiple 'R' and intercept in functional analysis of fish production at sample ponds

Particulars	Value
Intercept	2.8934
Multiple-R	0.57
Co-efficient of determination	0.32
F-value	13.50*
Sum of the elasticity	1.0874

* Significant at 1 per cent level of probability

The estimated multiple–R is about 0.57, indicating 57 per cent of fish production on ponds under study is explained by six variables included in analysis however the estimated co-efficient of determination shows 32 per cent of the total variation in fish production (y) explained by inputs which have been included in analysis.

The sum of production elasticities of individual resources indicates that nature of returns to scale, provided these resources are the only relevant inputs. In the present analysis, the variables included in analysis explain only 57 per cent of output hence the conclusion drawn on these variables may not be precise but an idea may be drawn on the basis of these variables. The sum of the elasticities of all the six variables under analysis is 1.09, indicating that the increase in 1 per cent in these variables may result in an increase of 1.09 percent in fish production. Hence, it may be said that the almost increasing return to scale exists in fish production on sample ponds.

Table 6.7 further revealed that the production elasticities of manures (X_1) fertilizer (X_2), lime (X_3), fingerlings (X_4) and feeds (X_5) are positive but the co-efficients of first four variables are significant at 1 per cent level of probability, whereas the co-efficient of feed is significant at 10 per cent level of probability. Hence, it may be said that there five variables have significant influence on increasing fish production in north Bihar. More specifically when other inputs remaining constant an increase in 1 per cent investment on manures is likely to increase fish production by 0.24 per cent. The similar conclusions may be drawn for fertilizers, lime, fingerlings and feeds. An increase in 1 per cent investment on any of these four inputs, keeping other inputs constant, the fish production is likely to increase by 0.24 per cent, 0.22 per cent, 0.27 per cent and 0.13 per cent, respectively. The comparatively higher expected increase in fish production has been noticed due to increase in investment of fingerlings because increase in investment on fingerlings may help fish farmers in using better quality fish seed which result in an higher increase in fish production. Production co-efficient of human labour is negative (-0.0140) but it is not statistically significant, indicating over utilization of human labour in fish production.

Table 6.7: Production co-efficients, standard errors and calculated 't' values of various input factors used in fish production

Variables	Production Elasticity	Standard Error	Calculated 't' value	MVP
Manures (X_1)	0.2442*	0.0700	3.487	5.9
Fertilizers (X_2)	0.2410*	0.0652	3.623	7.4
Lime (X_3)	0.2181*	0.0658	3.275	3.6
Fingerlings (X_4)	0.2665*	0.0751	3.546	3.4
Feeds (X_5)	0.1316**	0.0696	1.891	4.4
Human labour (X_6)	(-) 0.0140 NS	0.0793	0.176	-

*   Significant at 1 per cent level of probability
**   Significant at 10 per cent level of probability

On the basis of above analysis, it may be concluded that the fish farmers under study are not following the recommended package of practices of fish culture (Appendix-VI) since they are making much less investment on inputs like manures, fertilizers, fingerlings, lime and feeds but they are making more investment of human labours, probably due to abundance of family human labour force.

There is a need to augment the investment on inputs and reduce the investment on human labour for increasing fish production and making this enterprise most profitable. It could be done by imparting training to fish farmers of the project area which seems to be the weakest link in north Bihar, particularly in the field of fish production.

Marginal Value Productivities (MVPs) for all the five variables which have positive and significant production elasticities were estimated and it has been observed that all the variables namely; manures, fertilizers, time, fingerlings and feeds have MVPs of more than one, indicating that an investment of Re. 1 on either of inputs under analysis is likely to increase the fish production for more than the value of Re. 1 (Table 6.7.1). More specifically, an increase in investment of Re. 1 on manures may generate additional income of Rs. 5.7 from fish production. In similar way an additional investments of Rs. 1 on each of inputs namely fertilizers, lime, fingerlings and feeds are likely to generate additional income of Rs. 7.4, Rs. 3.6, Rs. 3.4 and Rs. 4.4, respectively through fish production. Hence, it may be said that there is a need to augment the investment in fish production in north Bihar for increasing fish production level and profitability to fish farmers which will ultimately go in long way in improving the quality of life of poor fishermen engaged in this enterprise.

Table 6.7.1: Marginal value productivities of inputs in fish production on ponds under investigation

Variables	Marginal value productivities (Rs.)
Manures(X_1)	5.9
Fertilizers(X_2)	7.4
Lime (X_3)	3.6
Finger lings (X_4) (fish seeds)	3.4
Feeds (X_5)	4.4

Marginal value productivity of human labour was not estimated due to negative and insignificant production elasticity.

Efficiency Measures

Profitability is the ultimate goal of any economic production process. Fish farmers put lot of labours and use various inputs to get high level of production and profitability. As discussed in the introduction section of this report, the average production is much lower in rivers, reservoirs and ox bow lakes and ponds in Bihar. But the fish ponds covered under Fish Farmers Development Agencies have performed better and produced more than 2100 kg of fish per hectare. But all the fish ponds of north Bihar are not covered under the project. Moreover, the steam of the agency has also been lost during last five years in Bihar. The low level of fish production is expected to generate low level of profitability to fish farmers in north Bihar.

However, profitability is one of the most important aspects in any economic activities for taking decision to invest in the particular economic activity/enterprise. A measuring stick is necessary to provide guide lines and standards for appraising accuracy of decision regarding the use of resources. One enterprise is said to be more efficient than the other when it yields a greater valuable output per unit of a valuable input. Various efficiency measures, therefore, need to be tried to express efficiency in fish production.

However, fish species wise conversion ratios has been discussed in the last sub-section to know the production potentiality of various fish species in North Bihar. In this section physical absolute measures namely; per hectare production, yield per kg of fingerlings, out put per day, Gross income, Net Income, family labour income, Farm business income and ratio measure namely return on investment are examined to know the level of profitability from fish production. Per hectare, per kg fingerlings, per 1000 fingerlings and per day out put of fish production on different size of ponds were estimated which are presented in Table. 6.8.

Table 6.8: Per hectare yield, yield per kg and per 1000 number of fingerlings and output per day on different categories of ponds

(Kg)

Particulars	Small	Medium	Large	Average
Yield (kg/ha)	1273.00	1601.00	1312.00	1416.00
Yield/kg of fingerlings	48.13	73.27	81.34	71.37
Yield/1000 no. of fingerlings	52.80	89.29	100.23	85.40
Output/day	831.00	8.06	5.72	7.50

It may be observed from the table that, on an average, per hectare of fish production was 1416 kg however it was comparatively higher on medium size ponds (1601 kilograms) followed by on large size ponds (1312 kilograms) and small ponds (1273 kilograms). Fish out put, on an average, per kg of fingerlings and per 1000 of finger lings were 71.37 kg and 85.40 kg, respectively, which declines with the decline in the size of ponds. This result clearly indicates that the large ponds owners either used quantity and number of fingerlings more closed to recommendation or stocked better quality of fingerlings. It has already been pointed out that the fish farmers stocked much higher quantity and number of fingerlings, particularly on smaller size of fish pond in north Bihar. It was mainly due to use of locally collected fingerlings from the rivers. Poor fish farmers operating small size of fish ponds neither had easy access to fish seed nursery located at distant places nor had funds to purchase better quality of fish seeds. Moreover they made comparatively higher investment on fish seeds which was only due to purchase of poor quality of seeds from local trader on credit basis.

On an average per day per hectare fish production was 7.50 kg but it was comparatively higher on small size of ponds (8.31 kg) followed by medium ponds (8.06 kg) and large ponds (5.72 kg). It was mainly due to late harvesting by large pond owners who harvest fish generally in the month of March–April whereas farmers of small pond owners used to start fish harvesting much earlier from month of January. It may be pointed out that fish farmers operating large ponds did not harvest fish at appropriate time which resulted in lower per day fish out put.

Per hectare gross income, net income, family labour income and farm business income were also estimated for fish production on ponds under investigation which are presented in Table 6.9.

Table 6.9: Per hectare Gross income, Net income, Family labour income and Farm business income on different categories of ponds

(Rs.)

Particulars	Small	Medium	Large	Average
Gross Income	36458.48 (28.64)	45124.83 (28.19)	37564.76 (28.63)	40286.41 (28.45)
Net income	14994.23 (11.78)	23566.54 (14.72)	21288.75 (16.23)	21223.47 (14.99)
Family labour income	20267.73 (15.92)	25774.40 (16.10)	21544.84 (16.42)	22969.76 (16.22)
Farm business income	25154.38 (19.76)	31326.49 (19.57)	25868.10 (19.72)	27845.77 (19.66)

Figures in parentheses indicate per kg income in respective category of ponds.

It may be observed from the table that the per hectare, on an average, gross income, net income, family labour income and farm business income were Rs. 40.29 thousand, Rs. 21.22 thousand, Rs. 22.77 thousand and Rs. 27.85 thousand, respectively in north Bihar.

Pond size wise analysis revealed that the medium size pond owners generated comparatively larger per hectare gross income, (Rs. 45.12 thousand), net income (Rs. 23.57 thousand), family labour income (Rs. 25.77 thousand) and farm business income (Rs. 31.33 thousand) than the corresponding income generated on large ponds (Rs. 37.56 thousand, Rs. 21.29 thousand, Rs. 21.54 thousand, and Rs. 25.87 thousand, respectively) and small ponds (Rs. 36.46, Rs. 14.99 thousand, Rs. 20.27 and Rs. 25.15 thousand, respectively).

It may further be observed that the difference in per hectare gross income of small and large ponds owners was about Rs. 1000 but the per hectare net income on small farm was about six thousand less than that of large farmers which was mainly due to the comparatively higher per hectare expenses in fish production on former than later size of ponds. It may be pointed out that the small pond owners employed a larger number of family labours in fish production (Table 3.11). It was the reason for very small difference in family labour income of small and large pond owners (Rs. 12 hundred only) since family labour income is sum of net income and imputed value of family labour wage.

The difference between small and large ponds also got narrowed down with respect to farm business income generated on these ponds because per hectare fixed investment was also higher on small ponds that of medium and large ponds.

Medium size pond owners emerged as most efficient in generating various types of income by utilizing input resources including family labour in probably more rational way than small and large ponds owners.

The ratio analysis would made this conclusion more reliable which is being discussed in succeeding pages.

The absolute measures of profitability indicate surplus income over expenses made in fish production. These figures may not give the true picture of the financial result of fish production with respect to investment. Hence, it would be more useful to further analyse these data to find out ratios between total investment and total out put (return to investment) forgetting an idea about per rupee return in fish production.

Total per hectare investment and return (gross income) and return on investment on three specified size of ponds were worked out which are presented in Table- 6.10.

Table 6.10: Return on investment on different categories of ponds

Particulars	Small	Medium	Large	Average
Total investment (Rs/ha)	21464.25	21558.29	16276.01	19062.94
Total return (Rs/ha)	36458.48	45124.83	37564.76	40286.41
Return on Investment (%)	69.86	109.32	130.80	111.33

It may be observed from the table that the return on investment, on an average, was estimated to 111.33 per cent on sample ponds but it was comparatively higher on large size of ponds (130.80 per cent) following by medium ponds (109.32 per cent) and small farm (69.86 per cent). All size of ponds generated much higher return on investment, indicating fish farming as a most profitable enterprise. Moreover, the comparatively higher return to investment on large size of ponds was only due to less investment in fish production. It may be noted that the traditional method of production by low investment generates the comparatively higher per rupee return than production through improved methods by investing more fund.

Functional analysis revealed that the sample ponds owners are operating in region-I. However, larger ponds owners seem to be operating much nearer to origin than that of other categories of ponds under investigation.

On the basis of above analysis it may be concluded that the fish farmers under investigation have generated much higher income from fish production but medium pond owners emerged as most efficient than the owners of other categories of ponds namely; large and small.

7
EXTENT AND PATTERN OF EMPLOYMENT

In India, an impressive growth has been achieved in rural sector during post independence period however the process got accelerated during green revolution period. But an important dimension of rural development is the spatial variations which is very relevant in a large country like India, having a wide range of socio-political, cultural and infrastructure bases. Despite the unprecedent growth in rural economy, as many as 40 per cent population is still in absolute poverty. A vast majority of them belong to labour community who continue to be poor due to lack of regular employment, low productivity, low income and higher family consumption. The occupational distribution of working population in India indicates that inspite of the phenomenal growth of non-agricultural sector in the last five decades of planned economic development there has been no significant change in the structure of the labour force engaged in primary sector. In India, production capacity is much below the needed level, which is, no doubt, increasing but with very slow rate. As against this, an addition to labour force is being made at a faster rate on account of the rapid growing population. Labour force in agriculture sector is rising annually by 40 lakhs, of which nearly 14 lakhs well in the rank of agricultural labour. It is said that a rapid development of industries can not absorb the rising labour force and it is more true in changed economic scenario hence the bulk of them will have to be provided with employment in the rural sector through intensive agriculture including draying, fisheries and other allied activities.

One of the most daunting challenges facing India is to provide employment not only for the additions to the labour force but also to reduce the back log of unemployment accumulated from the past. It is important to note that average annual growth rate of overall employment (in both organised and unorganised sectors) declined continuously from 2.75 per cent in seventies to 1.77 per cent in eighties but increased by 2.37 per cent in the nineties. The rapid increase in the rural labour force seems to be one of the important factors responsible for aggravating unemployment and consequently the rural poverty and violence in rural areas.

In Bihar, workforce increased from 17.24 millions in 1971 to 20.76 million in 1981 which further increased to 27.78 million in 1991. Out of an increase of 10.54 millions workforce during last 20 years, about 90 per cent of them got employment in primary sector, particularly in agricultural and agricultural allied activities. Number of workforce engaged in secondary sector declined from 1.51 millions in 1971 to 1.29 millions in 1991. The employment in tertiary sector increased marginally from 2.33 millions in 1971 to 3.61 millions in 1991. In clearly indicates the decomposition of secondary sector in Bihar. The tertiary sector also could not grow at desired level due to poor infrastructure.

It could be seen in the context of acceleration in the growth rate of the labour force as well as secular downward pressure on the employment intensity of the growth process. Economic growth and employment opportunities in themselves may not be sufficient to improve the living conditions of the poor. It is, therefore, necessary to shift the focus of employment strategies from crop production to allied agricultural activities like; fisheries, dairy, beekeeping, mushroom production etc. which may lead to better living and working conditions of rural labours. Estimates of rural fisheries labour force and their occupational characteristics are not available neither in the decennial population census nor in the national sample survey. In spite of the absence of any established data some projections have been attempted with a view to having a broad idea about the magnitude of the problem of rural fisheries labour force in future. Certain estimates at National level show that the share of fisheries labour has increased in total labour force during last 50 years. Number of fishermen has been estimated to 11.13 lakhs in Bihar, comprising 2.95 lakh full time, 4.88 part time and 3.30 lakh occasional. However, there is a dearth of information about their socio-economic status and employment pattern.

In the present section an effort has been made to examine their socio-economic status and employment pattern on the basis of data obtained from the villages under investigation.

Fishing households and population

Data were collected from 59 villages (18 clusters of villages) of Six districts of north Bihar. Per village number of fish farmer households, on an average, varied from one to 15 but per village number of fishery labours was worked out to be 29 which varied from 10 to 40 households in the study villages. Fishery worker engaged in fish farming was estimated to 6347 in study villages which has been worked out to be 42.90 per cent of total fishery worker population (Table 7.1) Male, female and child fishery workers constituted about 63.12 per cent, 39.63 per cent and 33.68 percent to their respective population. It clearly indicates the participation of all categories i.e. Male, female and children in fishery activities.

Table 7.1: Total number of households, fishery households, fisherman population and fish worker in study villages

Particulars	Number
Number of District	6
Total no. of study villages	59
Total no. fisherman and fishery labour households	1740
Total population of Fishery households	
Person	14795 (100.00)
Male	3945 (26.66)*
Female	3198 (21.61)
Children	7652 (51.72)
Worker engaged in fish farming	
Person	6347 (42.90)**
Male	2490 (63.12)
Female	1267 (39.63)
Children	2577 (33.68)

*   Percentage to total fisherman population
**   Percentage to total respective population

The detailed analysis of extent and pattern or employment is based on data collected from 180 fishery workers who were attached with the sample fishery households under investigation. Before analysis the extent and pattern of employment an attempt has been made to study the socio-economic status of fishery labours under investigation.

Age of fishery labour

Age is one of the important demographic characters which influences efficiency of workers in a particular activity.

Workers under investigation were categorised in four age groups, that is less than 25 years, 25-40 years, 40-55 years and 55 years and above (Table 7.2).

Table 7.2: Age group-wise distribution of fisheries labour

Age group line (Year)	No. of labours	Percentage
25-40	54	30.11
40-55	118	65.55
55 and above	8	4.44
Total	180	100.00

It may be observed from the table that the younger people (below25 years of age) did not participate in fishing activities and only 30.11 per cent of total fishery labours under study belonged to age group of 25-40 years. The majority of fishery workers (65.55%) belonged to the age group of40- 55 years and only 4.44 per cent to the age group of 55 years and above.

The above findings clearly indicate that the young workers did not take interest in fishing activities, probably due to strenuous jobs. Moreover, younger generation do not like to work in difficult situation since fishing work is generally practised in the deepwater.

Family size

An analysis of family size of fishery labours revealed that the majority of fishery labour households (53.33%) had large size family (8 and above members), 40.00 per cent, had medium size family (5–8 members)and only 6.66 per cent had nucleous family (Table 7.3).

Table 7.3: Distribution of fisheries labours in different size of family

No. of family members	No. of workers	Percentage
Less than 5	12	6.66
5-8	72	40.01
8 and above	96	53.33
Total	180	100.00

Hence, it may be said that the fishery labours had comparatively larger size of family than the average size of family in Bihar (5.2 members). It was mainly due to fact that the fishery labour households had to diversify their employment to crop production to meet food requirements which might promted them to keep extended family. A larger size of family has larger number of workers who could be employed in different fishing and non-fishing activities.

Occupation of fishery labours

Main and subsidiary occupations of fishery workers have been examined to know their extent of involvement in fishing activities. Main and subsidiary occupation of fishery labours were computed which are presented in Table 7.4

Table 7.4: Occupation-wise distribution of the respondents (Fishery labours)

Occupation	Main	Subsidiary
Fishery labours	176 (97.77)	4 (2.22)
Agril. Labour	-	170 (94.44)
Other	4 (2.22)	-
Total	180	174

Out of 180 fishery labour, 176 (97.77%) had fishing as main occupation and 4 fishing labour under investigation had labour in non farm sector as their main occupation, but we tried to analyse the subsidiary occupation of the respondents also. The table revealed that 97.77 per cent labours adopted fishing activities as their main occupation while only 2.22 per cent labours had adopted fishing as their subsidiary occupation. Those labours who had fishing as their subsidiary occupation were either land lord or businessman. The table also revealed that out of total labour, 94.44 per cent had agricultural labours as their subsidiary occupation. The majority of the fisheries labour engaged themselves in agricultural activities in lean period of crap farming.

About 2.22 per cent of fishery labours under study had main occupation other than agriculture and fishing. None of the fishing workers under study had agriculture as main occupation. However, 94.44 per cent of fishing workers had agriculture labour as subsidiary occupation and only 2.22 percent had fishing as subsidiary occupation.

It is worth pointing out that only 6, out of 180 fishing workers were dependent only on fishing activities and 170 fishing labours had agriculture as subsidiary occupation. It clearly indicates that the fishery do not provide required days of employment to workers engaged in fish production hence they had to engage themselves in other than fishing activities to meet family requirements. Hence, fishery sector needs to be developed in Bihar so that it could generate sufficient employment and income to workers engaged in this sector.

Access to Assets

All the fishery labours belonged to weaker section of society because all of them belonged to fishing communities(Mallah) who are known weaker section in north Bihar but only 12 per cent of them had agricultural land below 0.10 ha. Hence, all of them maybe considered as land less labour. Information on a dwelling houses revealed that 97.77 percent of them had own house and 2.23 per cent were residing in house built on landlord's land (Table 7.5). Hence, it may be said that 4 fisherman, out of 180 under study are still to have their own house and they are totally dependent on the sweet will of landlord to whom they are attached for fish production.

Table 7.5: Possession of Dwelling houses of the respondents

Types of dwelling house	No. of respondents	Percentage
Own house	176	97.77
Rented	-	-
Landlord house	4	2.23
Total	180	100.00

The type of dwelling houses possessed by fishing labours under investigation have also be analysed to know the living condition of fishery workers.

Out of 180 fishing workers, 16 (8.88%) had pucca house, 106 fishing workers (58.88%) had Katcha–thatched house, and 58 (32.24%) had Katch-Pucca thatched house. It clearly indicates that the majority of fishing labours are residing in either Katcha–thatched houses or Pucca-katch thatched houses. Fishery labours who are residing in Pucca houses, got their pucca houses constructed under Indira Awas Yojna. None of the fishery labours constructed Pucca houses through their earnings. The above findings clearly indicates poor living conditions of fishery workers in the project area.

An attempt has also been made to examine the possession of assets of fishery labour households. The major assets were catgorised in 6 groups that is; Dwelling house, livestock and Poultry, Fishing Equipments, Transport carrier (Cycle), Electronic asset, and Hand pipe. Per household average value of these equipments were worked out which are presented in Table 7.6.

Table 7.6:   Possession of assets on sample households

Assets	Average value in rupees	Percentage of total
Building	6747.22	45.0
Livestock and poultry	6030	40.2
Fishing equipments (Net, Kudal, Topiya)	616.55	4.1
Transport equipments Cycle	694	4.6
Electronic assets (Transistor)	497.66	3.3
Hand pipe	407.77	2.7
Total	14993.2	100

It may be observed from the table that on an average, per household value of total assets are estimated to Rs. 14.99 thousand. The value of dwelling house (except land) and livestock constituted about 85 per cent of the value of assets on fishery labour households.

Fishery labour households kept about 4.1 per cent of total asset in the form of fishing equipment namely net, kudal, topiya etc. Despite the high level of poverty, almost all the fishery labour households had cycle, watch and transistor. Handpipe for drinking water was also an important asset which was kept by about one-fifth of fishery workers under study. An average investment on this particular asset constituted about 2.7 per of the total value of assets owned by fishery labours under study.

Hence, it may be said that the living condition of fishery labours has been poor and all of them do not have access to safe drinking water. The majority of households have cycle, transistor and watches, indicating their exposure to outside world.

Extent of employment

The present section is devoted to examine the extent of

employment to fishery labours under investigation. The labourers were categorised on the basis of employment in five categories which are presented in Table 7.7. It may be observed from the table that majority of fisheries labours i.e. 41.1 percent could get employment in fishing activities for less than 100 days in a year, 26.6 per cent had 100 to 150 days employment in fisheries, 18.9 per cent got 150 to 200 days of employment and only 13.4 per cent had employment more than 200 days in a year. On the basis of the above observation it may be concluded that the majority of fishery labour could get employment of less than 150 days in a year when prompted them to search employment in other than fishery sector.

Table 7.7: Period-wise employment in fisheries activities

Employment days	No. of workers	Percentage
Less than 100 days	74	41.4
100-150	48	26.6
150-200	34	18.88
200-250	12	6.66
250 above	12	6.66
Total	180	100

As we had observed, 96.66 per cent fisheries labours had subsidiary occupation (Table 5.4). Agriculture emerged as an important subsidiary occupation of the fishery labours. Since 94.44 per cent of the household had agriculture as subsidiary occupation Table 7.8 revealed that the majority of fishery labour (67.9%) could get employment in agriculture upto 40 days however remaining 35.6 per cent of fishery labour got employment of more than 40 days in agriculture. Moreover, about 6 per cent of fishery labour could get employment in agriculture for more than 120 days in year.

Table 7.8: Period-wise employment of fishery labours in agriculture

Employment days	No. of workers	Percentage
Less than 40	46	64.4
40-80	26	14.44
80-120	22	12.2
120-160	8	4.5
160-200	2	1.1
Total	180	100.00

Employment Pattern

Monthly employment pattern of fishery labour has also been studied to know their pattern of employment. Month-wise average days of employment in fishery and agricultural activities of fishery labour under investigation were computed which are presented in Table 7.9 .

Table 7.9: Month-wise average employment of fishery labour in fishery and agriculture sector

(man days)

Month	Fishing	Agriculture	Total
June	8.3	3.6	11.9
July	16.8	6.2	23.0
August	9.8	3.1	12.9
September	2.4	2.0	4.4
October	2.0	2.4	4.4
November	7.4	6.0	13.4
December	9.6	7.0	16.6
January	12.6	1.8	14.4
February	14.9	1.4	16.3
March	15.3	1.4	16.7
April	12.2	5.2	17.4
May	15.7	2.9	18.6
Total	127.0	43.0	170.0

It may be observed from the table that Fishery labour were employed for 170 days in a year, that is; 127 days in fishery and 43 days in agriculture. Fishery labour could get comparatively larger days of employment in the month of July (23 days), followed by in the month of May (18.6 days), April (17.4 days), March (16.7 days) and December (16.6 days), September-October were the slack period of employment for fishery labours. Since they got less than 5 days of employment in these two months. Moreover, their monthly employment was less than 14 days in the months of August (12.9 days), November (13.4 days) and June (11.9 days).

The comparatively higher days of employment in fishery activities was available in the month of July (16.8 days) because peak activities of pre-stocking and stocking operations in fish ponds are practised in this month in the project area. Despite the peak period of fish operation, fishery labours were also employed for on an average, 6 days in the month of July in Agricultural operations, mainly due to transplanting of rice in this particular month. Fishery sector is also seasonal because it provides employment in only peak seasons like; stocking and harvesting period. The fish harvesting starts from December which continues upto the month of May however peak fish harvesting is practised in the period February – March.

It has further been observed that the fishery labours were employed in agriculture, particularly in transplanting of rice (July), harvesting of paddy and sowing of rabi crops (December) and harvesting of wheat (April). It is important to note that the fishery sector generates employment in January to March which is slack season for agricultural operation (Figure –I).

Hence, it may be said that the fishery labours are generally under-employed in the project area but fishery activities generated employment when there was no employment in agriculture sector. The development of fishery sector may generate supplementary employment in rural areas because these two sectors are not close competitors for utilising human labour.

Operation-wise employment

The work load of fishery labours in fishing activities have been grouped in seven categories namely; prestocking, stocking, input application, Chowkidari, pre-harvesting, harvesting and marketing. Average number of days per fishery labour in different activities were computed and have been presented in Table 7.10.

Table 7.10: Operationwise average annual employment fishery labour

(in days)

Operation	No. of Days	Percentage
Pond Construction	-	-
Minor Repair	-	-
Pre-stocking	19.8	15.59
Stocking	8.8	6.93
Input Application	4.7	3.70
Chowkidari	51.2	40.31
Pre-harvesting	13.5	10.63
Harvesting	19.1	15.04
Marketing	9.9	7.80
Total	117.2	100.00

It may be observed from the table that the Chowkidari provided nearly 51.23 days of employment in a year which accounted for 40.31 per cent of the total employment in fishery followed by pre stocking (19.8 days), harvesting (19.1 days), pre harvesting (13.5 days), stocking (8.8 days) and input application (4.7 days).

The comparatively higher working days in Chowkidari was mainly due to poor law and order situation in the project area. Fish ponds are prone to poisoning in earlier stage of production and theft in the latter stage. Pre-stocking and harvesting are also important operations because these two operations jointly provided about one-third of employment in fish production because earlier operation had many components like weeding, cleaning, netting, liming etc. whereas latter (harvesting) required more labours for collecting fish from ponds. Moreover, pre harvesting operation also provided about 13 days of employment for preparing ponds for harvesting operation. Other operations like; input application and stocking are not important with respect to employment to fishery labours but these operations require, no doubt, less number of labours but skilled labours.

On the ponds under study, attached fishery labours were not engaged for repair and construction of ponds because minor repairs were done either by family labours or contractual labours.

On the basis of above discussions, it may be concluded that the fishery sector does not provide adequate employment to labours engaged in this sector. Fishery labours are not only under employed but they get irregular employment due to seasonality of operation in fish production. The small size of pond is the main reason for low level of employment generation. Moreover, almost traditional method of fish production is also responsible for low level of employment generation in fishery sector.

8

FISH MARKETING

A well organised marketing system is most important for the growth of any economic enterprises. Fish being perishable, the problem is most acute. There is practically no definite structure of fish markets in the country (Pandey and Chaturvedi, 1984). It is more true for Bihar since the marketing infrastructure is one of the weakest in the country. The marketing systems so far developed in the state are according to needs of the local fish merchants and not in accordance with the need of fish producers scattered in the villages. It was envisaged that the Fish Farmers Development Agency would improve the marketing infrastructure and marketing process but it could not exert any influence on fish marketing in the state. The pre-harvest contractors and auctioning the ponds just before harvest are not common in Bihar. Market intermediaries namely; commission agents, wholesalers, retailers and vendors are operating in fish markets. When the intermediaries are more, the price spread is high and most of profit went to the intermediaries, leaving only labour charge for the producer (Ranchan, V, 1984).

In the present section, an effort has been made to analyse the fish marketing system in north Bihar. The analysis is based on primary data collected from 180 fish farmers, 18 wholesalers, 18 retailers and 18 venders.

Pandey, M.R. and Chaturvedi, G.K. (1984). Inland Fish Marketing, Inland Fishery Resources in India, (ed. Srivastava, U.K. and Vathsala, S.), Concept Publishing Company, New Delhi p. 564.

Ranchan, V. (1984). A General Review of the Freshwater Fish Culture in India, Inland Fishery Resources in India, (ed. Srivastava, U.K. and Vathsala, S.), Concept Publishing Company, New Delhi p. 1913-95.

Fish Production and Market arrivals

Zone wise fish production and arrival in Regulated markets are computed on the basis of data obtained from the Department of fishery, and Bihar State Agril. Marketing Board, Patna which are presented in Table 8.1.

Table 8.1: Production and arrival of fish in Regulated markets of North Bihar (1989-99) .

(Tonnes)

Zone	Total Production	Local Arrival Fish	Per cent of Arrival to Total Production
Zone-I	74640	1980	2.65
Zone-II	30700	1678	5.47
North Bihar	105340	3658	3.47

Source: Arrival data were obtained from Bihar State Agricultural Marketing Board, Patna.

It may be observed from the table that, on an average, the total fish production in north Bihar was 105 thousand tonnes and total average arrivals in major regulated markets located in north Bihar was 3.66 thousand tonnes during the later part of nineties which accounts for 3.47 per cent of total fish production. However, the proportion of fish arrival to total respective fish production was comparatively higher in Zone II (5.47%) than that of Zone I (2.65%).

In regulated markets, there were two sources of fish arrival that is local arrivals and arrivals from other state (Table 8.2). On an average, local fish arrivals constituted 60.88 per cent however the share of local arrivals was much higher in Zone II (93.63%) than that of Zone I (46.96%). On the other hand, arrival of fish in regulated markets of north Bihar from other states was comparatively higher in zone I (53.04%) than that of Zone I (6.37%).

The comparatively higher annual arrival from hinter land of regulated markets of Zone II was mainly due to comparatively higher annual per capita fish production in districts located in the zone (2.22 kg) than that of Zone I (2.00 kg). Moreover, the arrival of fish from other states (particularly from Andhra Pradesh) reaches to Patna market, which is the main distribution centre for fish in Bihar. Regulated markets of Zone II are distantly located from Patna, which maybe one of the important reasons for low fish arrival in regulated markets of Zone II.

Table 8.2: Zone-wise Fish arrivals from State and out side state in North Bihar

(Tonnes)

Zone	Local Arrival	Arrival from Other State	Total Arrival
Zone-I	1980.2 (46.96)	2236.1 (53.04)	4216.3 (100.00)
Zone-II	1677.8 (93.63)	114.1 (6.37)	1791.9 (100.00)
North Bihar	3658.0 (60.88)	2350.2 (39.12)	6008.2 (100.00)

Figures in Parentheses indicate percentage of respective total market arrival in Regulated Markets.

As mentioned in the methodology chapter the detailed analysis of fish marketing is based on primary data collected from farmer-respondents and market functionaries namely; wholesaler, retailer and vendor.

Disposal Pattern of Fish

The analysis is based on primary data obtained from fish farmers under investigation. Pond size wise fish production, consumption, wastage and sale were computed which are presented in Table 8.3.

Table 8.3: Production, Consumption, Wastage and sale of fish on different categories

(Kg)

Particulars	Small	Medium	Large	Total
Production	25764 (100.00)	83253 (100.00)	83940 (100.00)	192957 (100.00)
Consumption	326 (1.26)	1263 (1.52)	2173 (2.59)	3762 (1.95)
Wastage	122 (0.47)	876 (1.05)	1131 (1.35)	2129 (1.10)
Sale	25316 (98.27)	81114 (97.43)	80636 (96.06)	187066 (96.95)

Figures in parentheses indicate percentage of the total Production on respective categories of ponds.

It may be observed from the table that about 96.95 per cent of fish produced on sample fish ponds was sold either to market intermediaries or to consumers directly. The sale proportion to production increased to 97.43 per cent on medium ponds, which further increased to 98.27 percent on Small size of ponds. It was mainly due to low level of fish consumption on smaller size of ponds. About 1.26 per cent of fish produced on small ponds was consumed but the corresponding proportions were higher on medium ponds (1.52 per cent) and large ponds (2.59 per cent). The wastage of fish on pond level was comparatively lower in north Bihar since only 1.0 per cent of fish produced on ponds under investigation got wasted however the proportion of wastage was comparatively higher on large ponds (2.59 per cent) which declined to 1.52 per cent on medium ponds and 1.26 per cent on small ponds. On the basis of above discussion it may be said that the fish production is market oriented in north Bihar.

Transportation

Transportation of fish from pond to market is one of the important activities in fish marketing. Besides head load, five types of carriages are commonly used for fish transport in north Bihar. These are Jeep, Cycle, Horse Cart, Tempo (four wheeled riksha) and tractor. Number of farmers transported fish through various methods/carriage were computed which are presented in Table 8.4.

Table 8.4: Number of fish pond operators using different mode of transport in the project area.

(Number)

Types of Carriage/ Method	Small Pond Operators	Medium Pond Operators	Large Pond Operators	Total
Head Load	17 (25.0)	7 (10.1)	4 (9.3)	28 (15.6)
Cycle	12 (17.6)	4 (5.8)	3 (7.0)	19 (10.6)
Risksha	8 (11.8)	6 (8.7)	- (7.8)	14
Jeep	31 (45.6)	48 (69.6)	30 (69.8)	109 (60.6)
Tractor	-	4 (5.8)	6 (13.9)	10 (5.5)
	68 (100.00)	69 (100.00)	43 (100.00)	180 (100.00)

Figures in parentheses indicate percentage to respective total fish farmers.

Table 8.4 revealed that the Jeep was the most common carriage of fish transport for pond to market place because 60.6 per cent of farmers under study used this source of transport. The proportion of small pond operators used this mode of transport was comparatively lower (45.6 per cent) than medium and large pond owners (70 per cent).

About 15.6 per cent of fish farmers carried their fish to market place on head load however the proportion of farmers carried fish on head loads was comparatively higher in case of small pond. Operators (25.0 per cent) than medium (10.1%) and large pond (9.3%) operators.

Cycle was also an important mode of transportation in fish marketing. It was used by17.6 per cent of small pond operators. Large pond operators also transported fish by head load and cycle through their fish labours. This mode of transport was used by medium and large pond operators in terminal harvesting when harvested fish quantity was smaller in quantity.

In case of large quantum of harvesting, 4 medium pond operators and 6 large pond operators transported fish by tractors to respective district Market. None of the small pond operators used tractor for fish transportation whereas none of the large pond operators used riskha for transportation of fish for marketing purpose.

On the basis of above discussions, it may be said that the jeep is the most important mode for fish transportation however traditional method like; head load and cycle are still in use particularly for transporting small quantity of fish harvested on small pond operators of mainly weaker section of society.

Grading

Grading of fish was not common at farmers level. However, 15 per cent of small pond owners and 17 per cent of medium pond owners who sold fish direct to consumers graded fish for marketing purpose. Wholesalers under enquiry did not practise the operation of grading for fish marketing. All the retailers and vendors were found conducting grading of fish before selling to consumers. There were two common bases of fish grading that is; species and size. Retailers and vendors reported that they graded fish on the basis of species, categories and size.

Sale Pattern

The sale pattern of fish has been discussed in two broad categories that is; pond sale and market sale. Purchasers at pond level include workers, consumers, retailer and vendors whereas purchasers at market level include wholesalers, retailers and vendors. Sale of fish to different groups of purchasers at pod and market levels were computed which are presented in Table 8.5.

Table 8.5:   Fish sale pattern on different categories of ponds

(Kg)

Middlemen	Small	Medium	Large	All Ponds
(A) Pond level				
1. Retailer	1659	7866	7781	17306
	(6.55)	(9.70)	(9.65)	(9.25)
2. Vendor	9762	16428	14052	40242
	(38.56)	(20.25)	(17.42)	(21.51)
3. Worker	638	1195	1724	3557
	(2.52)	(1.47)	(2.14)	(1.90)
4. Consumer	6912	3253	1248	11413
	(27.30)	(4.01)	(1.55)	(6.11)
Sub-total	18971	28742	24805	72518
	(74.93)	(35.43)	(30.76)	(38.77)
(B) Market Level				
1. Wholesaler/Commission Agent	4708	45248	351269	101225
	(18.60)	(55.79)	(63.58)	(54.11)
2. Retailer	1637	7124	4562	13323
	(6.47)	(8.78)	(5.66)	(7.12)
Sub-total	6345	52372	55831	114548
	(25.07)	(64.57)	(69.24)	(61.23)
Ground Total (A+B)	25316	81114	80636	187066
	(100.00)	(100.00)	(100.00)	(100.00)

About 38.77 per cent of total fish marketed was sold at pond level and 61.23 per cent at market level however small pond owners sold about three-fourths of marketed fish at pond level but medium and large pond owners sold only one–third of marketed fish at pond level. At pond level, vendor was the major purchaser who purchased about 21.51 per cent of pond level marketed fish in north Bihar however vendors purchased about 48.56 per cent of pond level marketed fish on small size of ponds but the corresponding proportion was 20.25 per cent on medium ponds and 17.92 per cent on large ponds. Moreover, the quantity of fish sold to vendors on

different categories of ponds kept just reverse trend, that is; the higher quantity on large ponds and lower on small ponds. Retailers was the second important purchaser of fish at pond level who purchased about 9.25 per cent of marketed fish however retailers might have preferred to purchase fish from medium and large pond owners, mainly due to better quality of large size fish produced on these two categories of ponds. About 6.11 per cent of marketed fish was sold to consumers in north Bihar however the consumers preferred small size of ponds for purchasing fish, mainly frequency of harvesting on small ponds for getting fresh fish because. Moreover, the harvesting starts at early stage on small ponds which may be the main reason for higher proportion of fish sale to was more than other two categories of ponds under investigator consumers. Workers who also performed the job of vendor, purchased, on an average, about 1.90 per cent of marketed fish in north Bihar. All the three categories of ponds did not differ much with respect to proportion of fish sale to workers. However, the large pond owners sold the comparatively larger quantity to workers (1724 kg) than medium pond owners (1195 kg) and small ponds owners (638 kg). As mentioned earlier, 61.23 per cent of marketed fish was sold at market level. At market level wholesalers were the major purchaser who purchased, on an average, 54.11 per cent of marketed fish in the project area. Large pond owners sold about 63.58 per cent of marketed fish to wholesalers however the corresponding proportion was 55.79 per cent for medium ponds but small pond owners sold only 18.60 per cent of marketed fish to wholesalers because they sold about 75 per cent of marketed fish at pond level. Retailers were the second important purchaser at market level also who purchased, on an average, 7.12 per cent of marketed fish in north Bihar which varied from 5.66 per cent on large ponds to 6.47 per cent on small ponds and 8.78 per cent on medium ponds. It may be pointed out that the wholesalers dealing with fish do not operate from market yard, hence the arrival data recorded by officials of different regulated markets may not be precise and reliable.

On the basis of above discussions, it may be inferred that the small pond owners sell their fish mainly at pond level and large and medium pond owners at Market level. It may further be concluded that the small pond owners sell more than two-thirds of produce to vendor and consumers whereas medium and large ponds owners sell their more than fifty per cent of produce to wholesalers in the markets.

Marketing channel in fish marketing

The marketing of fish flows from different channels which can be classified in two sub-groups that is; marketing channels at pond level and at district level markets. There were seven main channels of fish marketing which are presented in diagram-1 and mentioned as below:

Channels of Fish Marketing

A.   Market Level

| Channel (P-W-R-C) | I | Producer-Wholesaler/Commission Agent-Retailers-Consumer |

Channel   I   Producer-Wholesaler/Commission Agent-Retailers-Consumer
(P-W-R-C)

Channel   II   Producer-Wholesalers/Commission Agent-Vendor-Consumer
(P-W-V-C)

Channel   III Producer-Wholesaler/Commission Agent-Consumer
(P-W-C)

Channel   IVA Producer-Retailer-Consumer
(P-R-C)

B.   Pond Level

Channel   IVB Producer-Retailer-Consumer
(P-R-C)

Channel   V   Producer-Vendor-Consumer
(P-V-C)

Channel   VI   Producer-Worker-Consumer
(P- W-R-C)

Channel   VII Producer-Consumer
(P-C)

The quantum of fish marketed through different seven channels were computed which are presented in Table 8.6.

Table 8.6:   Disposal of fish through different marketing channels at different levels   (in Kg)

Level of Selling	I P-W-R-C	II P-W-V-C	III P-W-C	IV P-R-C	V P-V-C	VI P-W-R-C	VII P-C	Total
Pond Sale	-	-	-	17306 (23.86)	40242 (55.49)	3557 (4.91)	11413 (15.74)	72518 (100.00)
Market Sale	77430 (67.60)	19885 (17.36)	3910 (3.41)	13323 (11.63)	-	-	- (100.00)	114548
Overall	77430 (41.39)	19885 (10.63)	3910 (2.09)	30629 (16.37)	40242 (21.52)	3557 (1.90)	11413 (6.10)	187066 (100.00)

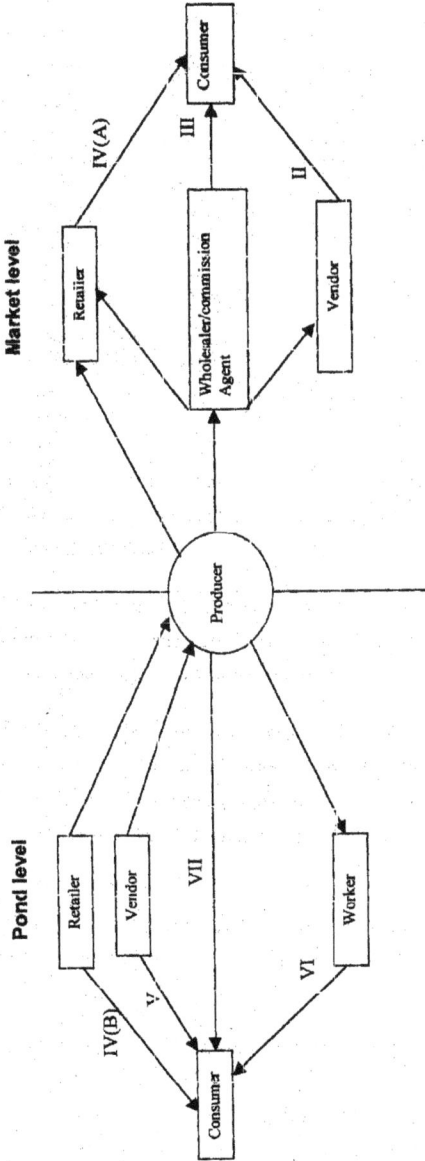

Diagram 1

Channels of Fish Marketing in North Bihar

It may be observed from the table that Channel I (P-W-R-C) was the most important channel for fish marketing in north Bihar since about 41.39 per cent of marketed fish was sold through this channel.

Channel V (P-V-C) was the second important channel which constituted about 21.52 percent of marketed fish on sample ponds. This channel was found only at market level because none of the producers sold fish to vendor at pond level.

Channel IV (P-R-C) was the third important channel but this channel was found popular in both the pond and market levels fish marketing which constituted about 16.37 percent of marketed fish on sample ponds however the sale through this channel constituted 23.86 per cent fish marketed at pond level and 11.63 percent at Market level.

Channel II (P-W-V-C) was the fourth important channel in the project area because 10.63 per cent of marketed fish on sample ponds was sold through this channel.

Channel VII (P–C), Channel III (P-W-C) and Channel VI (P-WR-C) were not common fish marketing channels since these three channels jointly constituted 10.09 per cent of marketed fish produced on sample ponds. The price spread in fish marketing has been examined through four important channels of the project area.

Price Spread

Price-spread is the difference between price paid by consumer and price received by producer. Due to presence of intermediaries in fish marketing it is often alleged that neither the producer gets remunerative price for his produce, nor the consumer gets the produce at a reasonable price for his consumption. The various studies were conducted on price spread in fish marketing and it was found that the farmer received about two-thirds of the price paid by the consumer, the rest was accounted for by the cost and profits of market intermediaries (Gupta, 1984).

To test this hypothesis, the price spread in four important channels have been examined which were worked out on the basis of data obtained from farmer respondents and market functionaries under investigation.

Data on price paid to producer, sale and purchase prices of different, market intermediaries including producers expenses incurred in marketing at all stages profit (margin) earned, and prices paid by consumers in four important marketing channels were computed which are presented in Table 8.7

Analysis of data revealed that the fish producers received per kg fish price of Rs. 35.80, Rs. 35.80, Rs. 31.20 and Rs. 24.80 in fish marketing through channel I, channel II, channel IV and channel V, which accounted for 88.31 percent, 87.53 per cent, 84.90 per cent and 81.31 per cent, respectively. Fish farmers made expenses on labour, transportation, packaging, commission to wholesales, unauthorized charges and misc.charges, particularly in channel I, channel II and channel IV which were amounting to Rs. 4.25/kg in each of channel I and channel II and Rs. 1.78/kg in channel IV.

If these costs are deducted from the producers' price, net producers' prices per kg of fish are worked out to be Rs. 31.55, Rs. 31.55, Rs. 29.42 and Rs. 24.90 in Channel I, Channel II, Channel IV and Channel V which accounted for 77.87 per cent, 77.14 per cent, 80.05 per cent and 81.64 per cent respective, of consumers price. In north Bihar, fish farmers are getting quite reasonable shares in consumers' rupee because the fish marketing channels are shorter due to sale of fish upto district level market. It was probably due to low level of fish production vis-a-viz demand for fish in the project area.

Table 8.7: Price spread in fish marketing through important marketing channels.

(Rs./Kg)

Particulars	Channel I (P-W-R-C-)	Channel II (P-W-V-C)	Channel IV (P-R-C)	Channel V (P-V-C)
(A) Producer's Price	33.80 (88.31)	35.80 (87.53)	31.20 (84.90)	24.80 (81.31)
(a) Expenses incurred by the producer sellers				
(1) Labour charge	0.16	0.16	0.10	0.10
(2) Transportation charge	0.85	0.85	0.48	-
(3) Packing charge	1.64	1.64	0.92	-

Contd.........

Gupta, G.S. (1984). Fresh water culture fisheries: Present Status, Prospects and Policy Issues, Indian fishery Resources in India (ed. Srivastava, U.K. and Vathsla, S.) Concept Publishing Company, New Delhi, 123-126p.

Particulars — Channel I	Channel II (P-W-R-C)	Channel IV (P-W-V-C)	Channel V (P-R-C)	(P-V-C)
(4) Commission Agent charge	1.04	1.04	-	-
(5) Unauthorised charge	0.32	0.32	0.18	-
(6) Wastage	0.10	0.10	0.18	-
(7) Other expenses	0.14	0.14	0.05	
(b) Well producers price				
(B) Actual purchase price of Commission Agent/wholesaler	31.55 (77.82)	31.55 (77.14)	29.42 (80.05)	24.90 (81.64)
(a) Expenses incurred by Commission Agent/Wholesaler				
(1) Labour charge	0.15	0.15	-	-
(2) Market fee	0.35	0.35	-	-
(3) Other expenses	0.06	0.06	-	-
(b) Profit margin of commission Agent/Wholesaler	1.18	1.18	-	-
(C) Sales price of Commission Agent/Wholesaler or purchase price of the retailer/vendor	37.54	36.50	-	-
A.(a) Expenses incurred by the retailer				
(1) Labour charge	0.22	-	0.16	-
(2) Transportation charge	0.04	-	0.10	-
(3) Market charge	0.25	-	0.20	-
(4) Wastage	0.17	-	0.18	-
(5) Other expenses (including unauthorised charge)	0.10	-	0.15	-
(b) Profit margin of retailer	2.22	-	4.76	-
B.(a) Expenses incurred by vendor				
(1) Vending cost	-	0.20	-	0.18
(2) Wastage	-	0.25	-	0.26
(b) Profit margin of vendor	-	3.95	-	5.16
(D) Sales price of retailers/vendors or Purchase price of the consumer	40.54 (100.00)	40.90 (100.00)	36.75 (100.00)	30.50 (100.00)

Figures in the parentheses indicate producer's share in the consumer rupee.

Note: Unauthorised charge taken by police and local Rangdar at almost all levels of intermediaries except commission agent level.

The difference in price received by producers in different channels was mainly due to difference in quality (species and size) of fish sold through different channels. Fish farmers sold better quality fish through Channel I and Channel II and inferior quality through Channel IV and Channel V.

Similarly, consumers paid, on an average, higher price in Channel I (Rs. 40.54/kg) and Channel II (Rs. 40.90/kg) than Channel IV (Rs. 36.75/kg) and Channel V (Rs. 30.50) due to difference in quality of fish.

Marketing cost

While analysing the fish marketing cost incurred in four marketing channels under investigation, it was found that per kg marketing cost was comparatively higher in channel I Rs. 5.59) followed by channel II (Rs. 5.26), channel IV (Rs. 2.57) and Channel V (Re. 0.54) (Table 8.8). Fish is a perishable commodity which needs proper packaging for marketing. Per kg cost of packaging ranged from Re. 0.92 in channel IV to Rs. 1.64 in each of channel I and channel II but its proportion to total cost of marketing was worked out to be 35.80 per cent in channel IV, 31.19 per cent in Channel II and 29.35 per cent in Channel I. There was no packaging cost in channel V because vendors purchased fish from fish farmers and sold directly to consumers, generally within 12 hrs of purchase.

Transportation cost was incurred in each channels under investigation and it was much less in amount which varied from Re. 0.18/kg in channel V to Re. 0.58/kg in Channel IV, Re. 0.89/kg in Channel I and Rs. 1.05/kg in Channel II but constituted 15.92 per cent, 19.96per cent, 22.57 per cent and 33.33 per cent of respective total marketing cost. Per kg transportation cost varied as per the geographical spread of the channel, that is the higher the geographical spread, higher the transportation cost. The cost incurred on payment of commission by farmers to commission agents and market fee by wholesalers to Agriculture produce Market Committee jointly amounted to Rs. 1.39 in Channel I and Channel II which constituted about one-fourth of total marketing cost in both the channels.

Market charges was an small amount paid by retailers to market contractors which was Re. 0.25/kg in Channel I and Re. 0.20/kg in Channel IV. Wastage was also an important cost in fish marketing but it ranged from Re. 0.23/kg in Channel IV to Re. 0.26/kg in Channel V, Re. 0.27/kg in Channel I and Re. 0.35/kg in Channel II.

Fish farmers paid unauthorized charges (deduction) by wholesalers at district level markets. Farmers paid Re. 0.32/kg in both the Channel I and channel II and Re. 0.18/kg in Channel IV. There was no unauthorized deduction in Channel V because fish was sold at pond level and none of the market functionaries except vendor was involved in this channel.

Table 8.8: Per kg cost incurred in fish marketing in important marketing channels

(Rs./Kg)

Particulars	Channel I (P-W-R-C-)	Channel II (P-W-V-C)	Channel IV (P-R-C)	Channel V (P-V-C)
Labour Charge	0.53 (9.48)	0.31 (5.89)	0.26 (10.12)	0.10 (18.52)
Transportation Charge	0.89 (15.92)	1.05 (19.96)	0.58 (22.57)	0.18 (33.33)
Packaging Charge	1.64 (29.35)	1.64 (31.19)	0.92 (35.80)	-
Commission and Market fee *	1.39 (24.86)	1.39 (26.43)	-	-
Market Charge*	0.25 (4.47)	-	0.20 (7.78)	-
Unauthorised Charge	0.32 (5.72)	0.32 (6.08)	0.18 (7.00)	-
Wastage	0.27 (4.83)	0.35 (6.65)	0.23 (8.95)	0.26 (48.15)
Other Charges	0.30 (5.37)	0.20 (3.80)	0.20	-
Total Marketing Cost	5.59 (100.00)	5.26 (100.00)	2.57 (100.00)	0.54 (100.00)

Figures in the parentheses indicate percentage to the total marketing cost.

* Market fee is collected by the respective Agricultural Produce Market Committee and Market Charges from contractors of Market/hat.

Profit Margins

Profit margin varied from channel to channel (Table 8.9). As expected wholesalers realised comparatively less amount of profit (Rs.1.18/kg) than retailers (Rs. 2.22/kg to Rs. 4.76/ kg) and vendors (Rs. 3.95 kg to Rs. 5.16 /kg).

Table 8.9: Marketing margin of intermediaries in important marketing channels

(Rs./Kg)

Intermediaries	Channel I (P-W-R-C-)	Channel II (P-W-V-C)	Channel IV (P-R-C)	Channel V (P-V-C)
Wholesaler/Commission Agent	1.18 (34.71)	1.18 (23.00)	–	–
Retailer	2.22 (65.29)	–	4.76 (100.00)	
Vendor	–	3.95 (77.00)	–	5.16 (100.00)
Total	3.40 (100.00)	5.13 (100.00)	4.76 (100.00)	5.16 (100.00)

Figures in the parentheses indicate share of intermediaries in total marketing margin in the respective channel.

Per kg profit margin in fish marketing varied from Rs. 3.40 kg in Channel I to Rs. 4.76/ kg in Channel IV, Rs. 5.13/ kg in Channel II and Rs. 5.16/kg in Channel V, Vendor's margin was comparatively higher than retailers' margin because vendors operated with smaller quantity of fish and made available fish to the door of consumers and realised comparatively higher profit margin in per kg sale of fish in north Bihar.

On the basis of analysis of secondary and primary data on fish marketing it may be said that the substantial quantity of fish is still marketed at pond level. The fish which are brought to district market for sale are marketed outside the market yard. Due to deficient in fish production (demand for fish exceeded supply of local produced fish), the arrivals from outside state was much higher, particularly in Zone I. Despite the unorganised fish marketing, fish farmers received more than three-fourths of consumers' price but it was only due to short marketing channels.

It has further been observed that the fish farmers were liable to pay unauthorized charges in fish marketing. It was only due to poor supervision and monitoring offish marketing activities in regulated markets.

The cost in fish marketing was about upto 15 per cent of consumers price, mainly due to packaging expenses and wastage in fish marketing process. The measures to improve the marketing system through enforcing regulation and advising wholesalers and farmers to operate from market yards of Agriculture Produce Market Committee will increase producers' share in consumers rupee. However, it will also have some influence on reducing consumers price of fish in north Bihar.

9

CONSTRAINTS IN FISH PRODUCTION

It was envisaged to identify the constraints in fish production on the basis of responses obtained from fish farmers, co-operative officials and market functionaries. The most important constraint of fish production is menace of floods which generally wash away the stocked fish seeds from ponds with flowing water. The constraints, other than flood have been analysised and presented in four broad sub-heads that is; Institutional, Technological, Economic and Social.

Institutional Constraints

The organizational structure of fishery cooperatives is federal with three tier system that is; State level (one), Regional level (4) and block level (465). About 40 thousand members are registered in fishery co-operatives in Bihar. There are four regional Co-operative unions namely; Tirhut Darbhanga Regional Fishery Co-operative Union, Muzaffarpur, Patna Regional Fishery Co-operative Union, Patna, Bhagalpur Regional Fishery Co-operative Union, Bhagalpur and Koshi Regional Fishery Co-operative Union, Purnea. All 18 fishery co-operatives of the project area form the sample of the present study. Out of 180 fish farmers under investigation, 107 were members of fishery co-operative society.

While conducting survey it has been observed that almost all the fishery co-operatives under investigation were dead. They were only engaged in facilitating leasing of fish ponds from Government to fish farmers. These co-operatives helped farmer members in leasing in ponds by them. About 16.6 per cent of fishery cooperatives under investigation arranged fish seeds but none of them managed inputs and marketing of fish.

The probing to co-operative officials revealed that the fishery co-operatives were engaged in only facilitating leasing of ponds from the Government. It has been reported that despite the government order of leasing out fish ponds for three years, the leasing of fish ponds was done for one year only in 45 per cent of cases. The leasing of ponds for shorter period has been the main reason for almost negligible investment in the development of fish ponds in the project area. The officials of co-operative suggested the lease period of 10 years which may encourage poor fish farmers to invest more in fish production, particularly in development of ponds.

The organization of fishery co-operatives at block level also emerged as major constraint in the functioning of Co-operatives which affects fish production adversely. There has been lack of co-ordination and rapport amongst officials of co-operative societies organized at block level. The most of problems are location specific and relate to particular village which were not considered seriously at block level co-operative society. While interviewing fish farmers, it was found that about 70 per cent co-operative members under investigation do not know about functioning of fishery co-operatives. None of fish farmers reported that they attended any meating of General body of fishery Co-operatives. It clearly indicates that the fish farmers are unaware of functioning of fishery co-operative society in the project area. These co-operatives seem to be on paper only because their members do not know about the functioning of this particular organization.

Technological Constraints

At first, an effort has been made to analyse the level of input use viz–a–viz recommendations there of. Per hectare recommended level of inputs, actual use level and gap were estimated which are presented in Table 9.1.

Table 9.1: Recommended level of input use, actual use level on sample fish pond and gap in input use

Major Inputs	Average Recommended level of use	Actual Level of Use	Gap (Number/kg)	Percentage
Fish Seed (number)	6000	16,589	(+) 10580	(+) 176.33
Manures (kg)	3000	385	(-) 2615	(-) 87.17
Fertilizers (in kg)				
Urea	135	35	(-) 280	(-) 88.89
SSP	180			
Lime (kg)	2000	(-) 1600	(-) 400	(-) 20.00
Feed (kg)	1800	45	(-) 1755	(-) 97.5

It may be observed from the lable that the fish farmers applied only 12.83 per cent of manures, 11.71 per cent of fertilizers and 80 per cent of lime and 2.5 per cent of feed recommended for Scientific fish culture. The gap between recommended and actual use of inputs ranged from 20 per cent in case manures to 97.5 per cent in case of feed. As mentioned in the cultural practices section, about three-forths farmers added lime, half of farmers used manures, less than one-third farmers applied chemical fertilizers and only 7 farmers applied feed to the fish ponds. It may be said that the level of use was not only much less than recommendation but these inputs were not used by all the farmers under investigation. On the basis of above observations, it may be said the low level of adoption of Scientific method of fish production is one of the important constraints in fish production in North Bihar.

Almost all the farmers do not aware of the recommended level of input use. None of the fish farmers under investigation reported about their any rapport with fishery extension officials. Hence, the poor fishery extension service emerged as most important constraint of fish production in north Bihar.

It may be pointed out that the fish farmers under investigation used about 176 times more fingerlings than the recommendation. It was mainly due to two reasons; (1) the size of fish seeds used was less than 3 cm against the standard fingerling size of 5 cm, and (ii) high mortality in fish seeds.

The use of small size of fish seeds was only due to unavailability of desired size of fish seeds because there was almost non-existence of well managed fish nurseries in north-Bihar. Annual fish try production in Bihar is about 335 million which is not even sufficient to stock one-third of ponds under fish production*. Almost all fish farmers (98 per cent) were neither aware of nor access to fish nursery in Bihar. Fish farmers stocked fish seeds which were easily available to them. It may be pointed out that about 50 per cent and fish farmers used Indigenous fish species for stocking which may be an important reason for low level of fish production.

During the year of investigation, there was no infestation of diseases and pest in fish ponds. Moreover, farmers are unware of

---

* Handbook of Fisheries Statistics, 1996, Ministry of Agriculture, Fisheries Division, Government of India, New Delhi: p. 151

diseases of fish and their control. In case of infestation they did not make any effort to use any toxicants/medicine. Moreover, one medium and three large pond operators used toxicants as prophylatic measures against diseases and pests of fish.

Economic Constraints

Low adoption level of scientific fish production may be due to non-availability of liquid money to poor fish farmers. The majority of fish farmers had fish production as main occupation and, one an average, per household net income through fish culture was Rs. 15705 however it was only Rs. 4498 in case of households who were operating small size of ponds. As mentioned earlier about 40 per cent farmers had an average pond size of 0.30 hectare hence their average per household net income through fish production was estimated to be less than Rs. 1500 only. Due to low level of income, fish farmers might have not able to invest in fish production. None of the inter viewed farmers reported to avail facilities of financial assistance the institutional agencies. Hence, poor access to institutional credit system is also an important constraint in fish production in north Bihar.

Despite a large number of regulated markets, the fish marketing is still unorganised and less than 10 per cent fish produced in north-Bihar are marketed in regulated markets. The major part of fish produced in north Bihar is sold either at pond level or in the local markets. This practice is most common in case of small producers who do not afford to transport the small quantity of fish to regulated markets. It results only to low price to small fish producers.

Social Constraint

In north Bihar, small pond owners harvest their fish in early stage, mainly due to poaching and Rangdari tax which are very common in north Bihar. These discourage farmers in making heavy investment. Poor law and order situation is responsible for these illegitimate activities. Fish farmers reported that they are also harassed by Govt. officials who are responsible for maintaining law and order situation.

Importants constraints in fish production are surmised as below:

1. Fish ponds are available on lease for short time period.
2. Fishery co-operatives are inactive.

3. Low level of input use.

4. Use of small size of fish seeds of high mortality rate.

5. Lack of fish – nurseries is north Bihar.

6. Poor access of farmers to fish nurseries.

7. Almost non-existence of fishery extension services.

8. Poor access offish farmers to institutional credit system.

9. Un-organised fish marketing.

10. Frequent poaching and harassment by Govt. official/local leaders.

10

SUMMARY AND CONCLUSION

The present study is envisaged to examine the various socio-economic aspects of fish farming enterprise however the specific objectives are to examine the cultural practices, costs, returns, human labour employment, marketing practices and constraints in fish production in north Bihar. The study is based mainly on primary data which were obtained from 180 fish farmers (10 fish farmers from each block) 180 fishermen (labours), 18 each of main market functionaries that is; wholesalers, retailers and vendors. The sample fish farmers (180) also formed the sample for analysing the fish marketing system. North Bihar is the project area for the study but 6-representative districts namely; Samastipur, East Champaran and Darbhanga in North West Alluvial Plains (Zone I); and Madhepura, Purnea and Katihar in North- East Alluvial Plains (Zone-II) were identified on the basis of water area and fish production for selection of representative blocks and villages. Three blocks from each of six identified districts and three-to-five villages from each of 18 identified blocks were selected for detailed investigation.

Fish farmer respondents were selected through stratified random sampling technique but sample fishermen (labours) who were attached/worked for longer period on the pond (s) of sample fish farmers that is; one each for one sample farmer/formed the sample for studying employment pattern of labours, engaged mainly in fish production. Market functionaries were selected randomly from the list of market functionaries, that is; 3 from each category of market functionaries (whole salers, retailers and vendors), in each of six district level markets under study, making sample size of 54 for market functionnaries.

While examining the socio-economic profile of fish farmers, it was noticed that about 7.22 per cent of fish farmers belonged to young age group of below 30 years, indicating that the younger group of fish farmers do not prefer to take up fish farming as profession in north Bihar. It may be probably due to more strenuous job of fishing since fish culture operations are generally done in deep water. The education level of fish farmer is directly related to the size of ponds, that is; larger the pond size operated by fish farmers, the higher the level of their education. In other words, it may be said that the larger the pond size, higher the level of economic status of pond owners and consequently the higher the level of their education. More-than one-third large pond owners had larger size of family but about half of small and medium pond owners had medium size of family (5-8 members).

Fishing caste (Mallah) had the major stake in fish farming in north Bihar. Upper castes and lower castes constituted about 14 per cent of fish farmers under investigation whereas OBCs constituted about one forth of fish farmers under study.

The majority of respondent fish farmers (76.67 per cent) had fish farming as main occupation and 21.11 per cent had fish farming as secondary occupation. All the small pond owners had fish farming either as main or subsidiary occupation, indicating their dependence on fish farming for their livelihood. On the other hand, about 56 per cent of large pond operators had fish farming as main occupation and 37 per cent had fish farming as secondary occupation. Hence, it may be said that the fish farming is an important occupation in north Bihar.

About 41 per cent of fish farmer respondents were landless and 39 percent had land holding of less than 0.8 ha. About 20 per cent of fish farmers under study had land holding of 0.8 ha and more. One-third of medium pond owners and two- thirds of small pond owners were land less. There was skewed distribution of land among land owning fish farmers. Hence, it may be inferred that the fish farming is an occupation of poor households in north Bihar. Fishery co-operative society is expected to play pivotal role in fishery development but the organisation is almost non-functional in north Bihar. About 60 per cent of fish farmer respondents were member of their respective fishery co-operatives but they did not get any facilities other than arrangement of ponds on lease from Government through co-operative.

Average size of fish ponds under investigation was 0.74 hectare however size of two-thirds of them was below 1 hectare. About 38 per cent ponds belonged to small categories (< 0.50 ha) but these ponds had average size of 0.30 hectare. One- third of fish pond belonged to large category (1 ha and more) and their average size was 1.47 hectare. Water depth of ponds varied from 1.0 to 3.40 meters in small ponds, 1.5 to 4.80 meters in medium ponds, and 1.8 to 5.10 meters in large ponds whereas almost all small and medium ponds got dried in summer season.

In north Bihar, fish production is still practised as traditional enterprise. Pre- stocking operation starts from the last week of May however it is not practised in ponds where stocking is done just after harvesting, particularly in late harvested ponds. Moreover, it was done in only 24 ponds , out of 202 ponds under investigation.

The stocking is practised from 2nd week of June to mid-September. The month of July is the peak stocking period in north Bihar . There are two important types of fish culture that is; indigenous carp culture and exotic crop culture however the majority of ponds were stocked by composite carp culture that is; indigenous and exotic. All the 202 sample ponds were stocked by indigenous major carps with or without exocic carps. Moreover, 14.36 per cent of sample ponds were also stocked by a specific type of indigenous fish, i.e. Mangur, Rohu among indigenous and common carp among exotic carps were most commonly grown species in north Bihar. These two species of fish constituted two-thirds of fish seeds stocked in the project area. Fish farmers did not use the fish seeds in desired (recommended) proportion due to their unawareness.

Application of lime is practised generally before the stocking operation in fish ponds but 23 per cent of sample ponds were sprayed by lime after stocking operation up to the month of December. As many as three- fourths of sample ponds were sprayed by lime however wide variation in per hectare use of lime (50 kg to 1000 kg) indficated poor knowledge of farmers about its recommended dose for fish production.

Mahua oil cake was most common toxicant which was used 2 to 5 weeks befor stocking in fish ponds to eradicate predatory fish , insects and frogs but it was used only on 4 ponds out of 202 ponds under study. Per hectare use of toxicants ranged from 817 kg to 1762 kg against recommendation of 2000- 2500 kg for scientific fish production.

Dung was the most commonly used manure for fish production in north Bihar which was added in as many as 55 per cent of ponds under investigation. Dung was applied in the majority of ponds during August- December at an interval of 1 month. The most common inorganic fertilizers used in the project area were urea and single super phosphate. Fertilizer were applied in 28 per cent of ponds during October- December. Urea and single super phosphate were utilised in 40.35 per cent fertilizer using fish ponds (57) in two to three split doses whereas only diammonium phosphate was used in 31.58 per cent of fertilizer using ponds and only urea was utilised in remaining 22 per cent ponds. In north Bihar, fish farmers started using supplementary feeds but it is still used on about 25 per cent of fish ponds. Mustard oil cake, groundnut oil cake and rice bran were important supplementary feeds used for fish production in the project area.

Fish harvesting is spread over whole of the year but mainly concentrated during months of December – May and number of harvesting ranged from 3 to 6. About 45 per cent of ponds under investigation were harvested during the period of March–May. The early harvesting was done either to meet family and crop cultivation expenses or in fear of poaching which is very common in north Bihar, mainly due to poor law and order situation.

Per hectare pond cost of fish production was Rs. 19.06 thousand, constituting 73.71 per cent variable and 26.29 per cent fixed costs. The comparatively higher per ha cost of fish production was observed on medium size of ponds (Rs. 21.50 thousand) followed by small ponds (21.46 thousand) and larger ponds (Rs. 21.50 thousand) Among the major variable costs incurred in fish production, the comparatively higher expenses was incurred on human labour (Rs. 5.77 thousand) followed by fish seeds (Rs. 3.84 thousand) and lime (Rs.1.28 thousand). Manure, fertiliser, toxicants and feed are still not important items of expenses in fish production in north Bihar. The expenses on packaging materials and marketing jointly constituted 5.35 per cent of total cost incurred in fish production in the project area.

Per kg cost of fish production was Rs. 13.46, constituting Rs. 3.34 fixed cost and Rs. 9.92 variable cost. Per kg cost of fish production, was comparatively higher on small ponds. (Rs. 16.86) followed by medium (Rs. 13.47) and large ponds (Rs. 12.41). Per kg fixed and variable cost also declined with the increase in size of fish ponds.

When per hectare pond and per kg cost of fish production were examined as per different cost concepts (Cost Al, Cost A2, Cost B and Cost C), it was found that the cost of fish production per hectare was comparatively higher on medium size of ponds than that of other two categories of ponds under investigation but the per kg cost of production was comparatively lower on medium size of ponds than that of other two categories of ponds namely; small and large ponds. It was mainly due to comparatively higher level of adoption of improved methods of fish production on medium size of ponds which resulted in higher production and consequently the lower per kg cost of production.

Cost on human labour and fish seeds jointed constituted about half of total cost and two-thirds of expenses incurred on variable costs in fish production in north Bihar. Per hectare use of human labour was estimated to 153.07 days, constituting 62.42 per cent hired labour and 37.58 per cent family labours. Per hectare employment of human family labour was observed to be higher on small ponds (240.83 days) followed by medium ponds (183.62 days) and large ponds (100.50 days). The comparatively higher employment of human labour on small size of ponds was mainly due to higher use of family labour because most of them (90 per cent) had fish production as main occupation and they probably did not have other opportunities of gainful employment due to poor resource base.

It has further been observed that the larger number of human labour was employed in watch and ward (73.51 days) followed by harvesting (28.08 days), input application (26.92 days), stocking (11.30 days), marketing (6.53 days), pre harvesting (4.70 days) and pre-stocking (2.02 days). The trend holds true on all size of ponds under investigation. It is worth pointing out that the "watch and ward" was the domain of family labour on small ponds but medium and large pond operators had to hire-in human labours for this particular operation.

Per hectare expenses on human labour in fish production was worked out to be Rs. 4955 however, on an average, farmers were to make payment of Rs. 3209 to hired labour and remaining Rs. 1776 was unpaid that is; inputed value of family labour.

The species mix in fish seeds and stocking rate are two important determinants in economics of pisiculture. The composite carp culture has been commonly practised in north Bihar. Per hectare 16.58 thousand of fish seeds, amounting 19.84 kg were stocked in north Bihar which constituted 77.47 per cent of indigenous and 22.53 per cent of exotic fish species.

Among indigenous fish species, Rohu was the most popular carp which constituted 30.04 percent of total indigenous fish seeds used in ponds for stocking whereas the common carp was the most popular fish species among exotic fish species which constituted 40.72 per cent of exotic fish species used for stocking in north Bihar. Mangur, an indigenous species of fish has recently been introduced in East Champaran district of north Bihar but it was used for stocking, particularly on medium size of ponds.

It has further been observed that the proportion of rohu-the middle level species of fish to total stocked fish seeds was 28 per cent which was much higher than the prescribed proportion of 10-20 per cent for composite fish culture. The proportion of upper level fish (Catla + silver carp) and grass eating fish (grass carp) were 30 .78 per cent and 8.32 per cent, respectively which almost matched the recommended proportion of 30-40 per cent and 5- 15 per cent, respectively. The proportion of lower level fish (common carp + mrigal) was worked out to be 29.49 per cent against the recommended proportion of 40-45 per cent.

While analysing the sources of fish seeds, it has been observed that the farmers utilized multi- sources for procuring fish seeds but fish seeds traders emerged as the most important source since 78.89 per cent farmers purchased indigenous major carps, 12.22 percent fish farmers procured Mangur, and 77.22 per cent farmers procured exotic species of fish from this particular source. Fish co- operative was not an important source of fish seeds since the organization was almost dead in north Bihar. pond delivery of fish seeds was common practice however fish farmers had to procure Mangur species of fish from local and out side state markets. Moreover, one-fourth of farmers could get mangur fish seeds either on ponds or in the village itself.

Fish production, conversion ratio and potential production in north Bihar have also been examined. Per hectare fish production was worked out to be 14.16 quintals, constituting 73.94 per cent indigenous and 26.06 per cent exotic fish in the project area. The comparatively higher per hectare fish productivity was observed on Medium ponds (16.01 quintals) followed by large ponds (13.12 quintals) and small ponds (12.73 quintals). Per hectare higher fish production on medium ponds was probably due to use of comparatively higher quantity of fertilisers, lime and supplementary feeds on this category of ponds than that of other categories of ponds under investigation.

On an average conversion ratio was 71.37 however it was comparatively higher in case of exotic fish (82.55) than indigenous fish (68.12). Among indigenous fish, the comparatively higher conversion ratio was observed in case of Rohu species of fish (68.05) whereas as silver carp of exotic species exhibited higher conversion ratio (94.00). Despite the higher conversion ratio of silver carp, it was not preferred fish species for production in north Bihar, mainly due to inferior taste to other cateries of indigenous and exotic fish species.

Per hectare productivity of 20 quintals and more on 25.32 per cent of medium ponds and 10.53 per cent of small ponds clearly indicate the potentiality of fish production in the project area. Hence, there is a need to make concerted efforts in making the resources and technology available to fish farmers for increasing fish production in the state.

Production efficiencies of different factors of fish production were estimated by using cobb-douglas function model of regression. The cross sectional data of per hectare value of fish out put and value of major inputs namely; manures, fertilizers, lime, fish seeds, feed and wage of human labour were used for analysis. The estimated F-value (13.50) suggests that the independent variables included in analysis affect the fish production significantly on ponds under investigation. Analysis revealed that about 57 per cent of fish production on ponds under study is explained by six variables included in the analysis however the estimated co-efficient of determination shows 32 percent of the total variation in fish production explained by inputs included in analysis.

Analysis further revealed that the almost increasing return to scale exists in fish production on sample ponds. Manure, fertilizer, lime, fish seeds, and feeds have positie and significant coefficients, indicating that these variables had significant influence on fish production But production co-efficient of human labour is negative but not statistically significant. The estimated MVPs of manures, fertilizer, lime, fish seeds, and feeds clearly indicate that one rupee investment on each of these variables may generate additional fish out put of Rs. 5.9, Rs. 7.4 Rs. 3.6, Rs. 3.4 and Rs. 4.4, respectively.

Despite the low level of fish production, per hectare gross income was worked out to be Rs. 40.29 thousand however it was comparatively higher on medium size ponds (Rs. 45.12 thousand) than that of large (Rs. 37.56 thousand) and small ponds (Rs. 36.46 thousand). Net income, family labour income and farmbusiness income had the similar trend on different categories at fish ponds. It has further been observed that small and large ponds operators did not differ much with respect to generation of family labour income, probably due to higher per hectare employment of family labours and own capital in fish production.

Return to investment was estimated to 111 per cent however it was higher on large ponds (1340.8%) which declined with the decline in the size of ponds that is; 109.32 per cent on medium ponds and 69.86 per cent on small ponds. The comparatively higher return to investment on large ponds was probably due to low level of investment on fish production on these ponds.

The extent and pattern of employment of main fishery labours have also been examined. About 97.77 per cent of labours under study had earned their livelihood by working as fishery labours on other's ponds. These labours had poor asset base (Rs. 14.99 thousand) and inhygeinic living condition. Two- thirds of fishery labour could get annual employment of less than 150 days in fish production and four- fifths of these labours could get annual employment of less 80 days in crop sector. Fishery labours were, on an average, engaged for 170 days in a year, that is 127 days in fish production and 43 days in agriculture. In fish production they were employed for 51 days for *Chaukidary* however harvesting also provided employment of 32.6 days in a year, that is 19.1 days in harvesting and 13.5 days in pre- harvesting operation. Hence, it may be said that the fishery lalbours are under- employed and they get irregular employment in fishery and agriculture sector.

Fish produced in north Bihar is sold upto district level. Small pond owners generally sold fish at pond level. Farmers who brought fish to district level market paid unauthorised charges to wholesalers.

Farmers did not grade their fish for marketing purpose. There were four important fish marketing channels. About 90 per cent of marketed fish on pond under study were sold through these four channels. Producers, on an average, obtained three- fourths of consumers' price which seems to be reasonable but it was only due to shorter marketing channel. Per kg cost of packaging and wastage was substantial. Market fee was paid in two channels only because other two channels were operated outside regulated market. Hence, it may be said that fish marketing is still unorganised in north Bihar.

While analysing the constraints in fish production, menace of flood emerged as the most important constraint in fish production in north Bihar. However, other important constraints are as follows.

1. Short period of leasing pond
2. Degraded micro- environment of ponds, particularly, quality of water and siltation in the pond.
3. Primary fish co- operative is organised at block level which is an inactive organization.
4. Low level of input use in fish production.
5. Use of small size of fish seeds of high mortality.
6. Almost non- exsistence of fish nurseries in north Bihar.
7. Lack of access of poor farmers to fish nurseries.
8. Unavailability of Institutional credit for fish production.
9. Unorganized system of fish marketing.
10. Almost non- existence of fishery extension services.
11. Frequent poaching and harassment by Govt. officials and local leaders.

Conclusions

1. Fish production is the domain of poor farmers in north Bihar. The majority of them belong to socio-economically backward community (Mallah). Any improvement in fish production practices may likely to increase the income level and quality of life of these poor farmers.

2. Fish production is practised mainly on small size of ponds in north Bihar. Almost all the ponds are natural and none of them constructed during last 50 years. These ponds

have not been repaired/ desilted during recent years. These ponds are generally got dry in summer season. The majority of fish ponds are under lease arrangement for short period which is also an important factor for not making any expenses on repair or/and improvement of ponds. An increase in lease period of 10 years may encourage fish farmers to make necessary investment on repair/ improvement of ponds.

3. Fish production is labour intensive enterprise since it is still practised with traditional technology. There is a lack of awarness about improved package of practices of fish production in the project area, mainly due to almost non-existence of fish extension services in north Bihar.

4. Economics of fish production is favourable however it could further be improved if farmers are advised to adopt scientific package of practices

5. There is a shortage of fish seeds due to lack of fish nurseries in north Bihar.

6. Fish productivity is, no doubt, quite low but per hectare productivity of 2000 kg is quite common, mainly on medium size ponds with use of about half of the recommended level of inputs. It clearly indicates that there exists fish production potential which could be harnessed by educating farmers and making resources available to them for using recommended quantum of inputs in fish production.

7. Fishery co- operative is an inactive organization, mainly due to poor rapport amongst members.

8. Fish marketing is still unorganised. Farmers are required to pay unauthorised charges in fish marketing. Regulation of fish marketing is still to be implemented effectively in north Bihar.

Emerging Policy Issues

1. Arrangement should be made to lease out government fish ponds for longer period that is; at least five years and more.

2. Fishery development department needs to be reorganized to take more responsibilities to have better rapport with fish farmers. Institutional efforts should also be made to

train fish farmers to enrich their knowledge in the field of scientific fish production. NGOS may also be involved in imparting training to fish farmers.

3. Fish nurseries should be established at large scales to bridge the gap between supply of and demand for fish seeds in north Bihar.

4. To overcome the problems in fish production due to recurrent floods, fish farmers should be advised to adopt pen fish culture.

5. Credit Institutions may be advised to increase the credit flow for construction of fish ponds which will help increasing the employment opportunities to fishery labours who are under – employed in north Bihar.

6. Fishery co- operatives needs to be established at panchyat level so that the member could have close rapport among themselves. The revitalization of fishery co-operative will go in long way in improving the economy of fish farmers through arranging marketing of inputs and output of fish production system.

7. Fish farmers should be encouraged to bring their produce in market yards. It could be only done if wholesalers are advised to operate from market yards. The market regulation should also been enforced in such a way so that unauthorised deductions are not charged from fish farmers.

Appendix I

Fish Production and Export During 1960-61 to 1999-2000

Year	Fish Production ('000 tonnes)	Fish export (000' tonnes)	% of Fish export to Fish production
1960-61	1160	19.9	1.72
1970-71	1756	32.6	1.86
1980-81	2442	69.4	2.84
1990-91	3836	158.9	4.14
1995-96	4949	310.1	6.27
1996-97	5348	394.5	7.38
1997-98	5388	398.2	7.39
1998-99	5262	311.2	5.91
1999-2000	5056	390.6	6.90

Appendix II(A)

Classification Aquaculture System

Criteria	Kind
Purpose of culture	Human food
	Improvement of natural stock
	Sports and recreation
	Ornamental fish
	Bait
	Industrial products
Nature of enclosure	Pond culture
	Gage and pen culture
	Receway culture
	Raft culture
	Closed high-density culture
	Sea ranching
Sources of fry	Natural waters
	Captured gravid females
	Hatching
Level of management intensity	Extensive
	Semi-intensive
	Intensive
Number of species stocked	Monoculture (Single species)
	Polyculture (more than one species)
Water salinity	Fresh water
	Brackish water
Water movement	Running water
	Standing water
Water temperature	Cold water
	Warm water

Appendix II (A) contd...

Criteria	Kind
Food habit	Herbivorous species culture
	Omnivorous species culture
	Carnivorous species culture
Combination with	Rice-fish farming,
agriculture production	Poultry-fish farming
	Pig-fish farming

Appendix II (B)

Important Fish Available in Bihar

1. *Cat Fishes:* These include fish like Boari (Wallage attue) Ari (Mystusaor), Beanwa.

2. *Herrings:* Sardines and Hilsa only Hilsa which is primarily a sea fish used to migrate upstream in the Ganga for breeding.

3. *Carps:* These includes Catla, Rohu and Mrigal. These are cultured widely in tanks and ponds but are also found in the rivers.

4. *Anchovics*: These include Brinds or Dhanga and are found in the Ganga.

5. *Jew Fishes:* Garai in Paddy fields, river etc.

6. Murrels and Parchase (air Breathining fish) these include Kabai, Murrel, Mangur and Singhi. They are found in fresh water and occur mainly in North Bihar. Some of these fish are snake headed and some can climb. They constitute nearly 15 per cent of marketed inland fish and thrive well in Swamps and mans.

7. *Prawns:* Locally known as Jhinga.

8. *Crabs and Snails:* Crabs and Snails also occur mainly in Swampy area.

9. *Miscellaneous Fishes:* Setipinna phasa (Ham) Gadusia Chapra (Ham), Eutropiichthys vacha (Hem) Mystus (Osteobagrue) Aor (Ham) Seenghala (Syhes) Maorobrachium malcolsoni, Mastacembelus sps. Barilius sps. Lodyochilus, Mc clelland and Botia sps. Channas (Murrels) Bachwa (Eutropichthis vacha), Paswa (Setipinne phasa).

Appendix III

Districts of Zone-I and Zone-II

Zone-I: Samastipur, Darbhanga, Saran, Siwan. Gopalganj, East Champaran, West Champaran. Muzaffarpur, Vaishali, Sitamarhi, Madhubani, Begusarai.

Zone-II: Purnea, Saharsa, Katihar, Supaul, Madhepura, Khagaria, Araria, Kishanganj, Naungachia of Bhagalpur district.

Appendix IV

Average size of fingerlings utilized on sample ponds alongwith the range of size

(Size in cm)

Species	Average Size	Range of Size
Rohu	1.50	1 - 2.50
Catla	1.50	1 - 2.50
Mrigal	1.50	1 - 2.50
Common Carp	1.75	1.50 - 2.00
Silver Carp	1.75	1.50 - 2.00
Grass Carp	2.50	1.50 - 3.50
Others	2.00	1.50 - 2.50

Appendix V

Per hectare utilization of important factors of production on different categories of ponds

(quantity in qt.)

Particulars	Small	Medium	Large	All Ponds
Manures	5.05	4.47	2.99	3.85
Fertilizers	0.26	0.39	0.32	0.34
Lime	1.59	2.48	2.04	2.14
Fingerlings	0.26	0.22	0.16	0.20
Supplementary Feeds	0.37	0.51	0.42	0.45
Human Labour (Mandays)	240.82	183.62	100.49	153.07

Appendix VI

Per hectare recommended level of input in fish production

	Items	Quantity
1.	Mahua Cake(in kg)	2000-2500
2.	Lime (in kg)	500-2000
3.	Dung (in tonnes)	10 – 12
4.	Urea (in kg)	180
5.	Single super phosphate (in kg)	240
6.	Stocking rate of fingerlings (number)	5000 to 6000
7.	Supplementary feed	2-3 per cent of estimated weight of fish

BIBLIOGRAPHY

_____(1996). Development of Fisheries in Bihar, Situation, Progress, Problems, Potentiality and Strategy, Department of Institutional Finance and Programme Implementation, (Pariyojna Sangathan), Govt. of Bihar, Patna.

_____Govt. of Bihar (1991). Bihar At a Glance (1991). Deptt. of Statistics and Evaluation, Bihar, Patna.

_____Govt. of India (1982). Hand Book of Fisheries Statistics, Ministry of Agriculture, New Delhi.

Agbagani, R.F.; Ballao, D.D.; Franco, N.M.; Ticar, R.B. and Guanzon, N.G. (1989). "An economic analysis of the nodular pond system of milk fish production in Philippines. *Aquaculture*. 83 (3/4) :249-59.

Ahmad, S.H. and Singh, A.K. (1992). Present status Potentialities and strategies for Development of reservoir fisheries in Bihar. *Fishing Chines*. 12(8) : 49-57.

Ameen, M. (1987). " Fisheries resources and opportunities in fresh water fish culture in Bangladesh. *Noakhali Bangladesh*. 20 (1) :244.

Anil, M.; Desai, N.R.M. and Naik, B. (1996). Inland fishery Production in Karanataka State : An Economic Study in Simoga District. *J. Inland Fish Soco. India*. 28 (1): 7-13.

Asian Productivity Organisation (1989). "Fish marketing in Asia and the pacific". Report of a study mission, 1987. Japan : 182.

Bell, F.W. and E.R. Canterbery (1976). Acquaculture for the developing countries; A feasibility Study Cambridge, Mass Bellinger Publishing Co.

Bohle, H.G. (1985). "Towards new concepts in marketing geography" Proc. Of the International Conf. of the IGU study group. Gorakhpur, Concept Publishing Company, New Delhi (1988) : 417-25.

Chaudhary, S. and Rao, P.S. (1991). Economics of CARP SPAWN through Eco-Hatchery in Tripura, Newsletter – Indian Society for Fisheries Economics and Development, 1(2) : 5.

Chauhan, S.K.; Sharma, R.K. and Moorti, T.V. (1989). "The role of cooperatives in Himachal Pradesh Fish marketing". *Indian J. Agril. Marketing.* 3(2): 100-6.

Chauhan, S.K.; Vashist, G.D. and Moorti, T.V. (1989). "Economics of reservoir fisheries – a study of fish cooperatives in Poog dam area of H.P." *Indian Cooperative Review.* 26 (3): 329-37.

CIFRI(1979). "Operational economics of different culture system". Background information on research work and accomplishment of CIFRI, Barrackpore, West Bengal : 16.

Dekadrai, P.V. (1996). Growth in Fisheries and Acquaculture Resources and Strategies, "National Seminar on Agricultural Development Perspectives for the Ninth Five Year Plan, CMAIIM, Ahmedabad, June13-15, 1996.

Dhondyal, S.P. and Singh, G.N. (1966). "Benefit – cost ratio in fish culture". *Indian J. of Agril. Econ.*23 (4) : 237-39.

Dutta, A.K.; Sengupta, K.K. and Patra, S. (1978). "Composite fish culture". Technical Report on the progress of work and achievements at West Bengal centres during 1977-78, CIFRI, Brrckpore : 9.

Gangopadhyay, S. and Giri, A.K. (1990). "Augmentation of income through inland fish production – a case study". *Economic Affairs.* 35 (3) : 163-71.

Ghorai, A.K. (1979). Attainments of composite fish culture demonstration centres in W.B. during 1976-77". Bulletin No. 30, CIFRI.

Government of India (1976). Report of the National Commission on Agriculture, Part-IV- Fisheries, Ministry of Agriculture and Irrigation, New Delhi.

Government of India (1996). Handbook on fisheries Statistics, Ministry of Agriculture, Fisheries Division, New Delhi, 53-55.

Government of India (2000). Agricultural Statistics at A Glance, Directorate of Economics and Statistics, Ministry of Agriculture, New Delhi, 135-36.

Govt. of Bihar (1995). *Bihar Ek Jhalak,* Deptt. of Statistics and Evaluation, Patna (Bihar).

Gupta, G.S. (1984). Fresh water culture fisheries : Present Status, Prospects and Policy Issues, Indian fishery Resources in India(ed. Srivastava, U.K. and Vathsla, S.) Concept Publishing Company, New Delhi-123-126.

Gupta, G.S. (1984). Fresh water culture Fisheries, Present Status, Prospects and Policy issues, *Indian fishery Resources in India,* (ed. Srivastava, U.K. and Vathsala, S.) Concept Publishing Company, New Delhi: 123-156.

Jha, B.C. and Chandra, K. (1997). Kusheshwar Sthan Chaur (North Bihar), Central Inland Capture Fisheries Research Institute, Barrackpore, West Bengal : 1-15.

Keenum, M.E. and Waldrop, J.E.(1988). "Economic analysis of farms raised catfish production in Mississippi. *Technical Bulletin.* No. 155 : 27.

Khan, H.A.; Mishra, D.N. and Tyagi, B.C. (1979). "Investigations on composite fish culture with or without supplementary feeding of fish and fertilization. "Symposium on inland aquaculture" held at Barrackpore.

Kumar, D. (1996). Aquaculture Extension Services Review, FAO Circular No. 906, Food and Agriculture Organization of the United Nations, Rome.

May and Baker (1983). "Research highlights –sulphadizine to cure bacterial disease". *CIFRI Newsletters.* 6 (2) : 6.

Memon, K.N. (1996). Fishery in Vallebhsagar Reservoir, Ukai. *Fishing Chimes.* 15 (10) : 37-38.

Mishra, D.C. and Rao, P.L.N. (1979). "Different aspects of fish culture". Technical Report on the progress of work and achievements at Orissa Centres during 1975-78. Rural Agricultural Project, CIFRI, Barrackpore : 12-13.

Mishra, D.N.; Khan, H.A. and Tyagi, B.C. (1979). "Composite fish culture without fertilization and feeding". *Annual Report, CIFRI,* Barrckpore : 139.

Mollah, A.R.; Cowdury, S.N.I. and Habib, A.M. (1990). "Input-out put relations in fish production under various ponds size, ownership pattern and constraints". *Bangladesh J. Training* and *Development.* 3 (2): 87-101.

National Commission of Agriculture (1976). Ministry of Agriculture and Irrigation, Govt. of India, Vol. VIII : 27-41.

Pandey, A.K. and Qureshi, S.A. (1996). "Development of Fisheries in Sondur Reservoir". *Fishing Chimes.* 15 (10): 23-24.

Pandey, M.R. and Chaturvedi, G.K. (1984). Inland Fish Marketing, Inland Fishery Resources in India, (ed. Srivastava, U.K. and Vathsala, S.), Concept Publishing Company, New Delhi p.564.

Ranchan, V. (1984). A General Review of the Freshwater Fish culture in India, Inland Fishery Resources in India (ed. Srivastava UK and Vathsala, S.) Concept Publishing Company, New Delhi. P.1913-95

Randhir, M. (1977). "Economic of composite fish culture technology in India" *Annual Report* of CIFRI, Barrackpore : 42.

Randhir, M. (1986). "Economic investigation on carp culture and air breathing fish culture in India". *Annual Report,* CIFRI: 89.

Rao, D.V.S. and Chowdhry, K.P. (1988). "A Study of marketing costs margins and factors influencing prices of inland fish in selected markets. *Indian J. Agricul. Marketing.* 2 (2): 170-75.

Rao, J.K. and Rama Raju, T.S. (1989). "Observations on Polyculture of carps in large freshwater ponds of Kollerulake". *J. of Aquaculture in Tropics.* 4(2) : 157-164.

Rao, K.R.M. and Raju, R. (1986). Distribution of fisheries in Andhra Pradesh : A study. *Indian J. Marketing.* 27 (4): 17.

Rao, P.S. (1968). Some aspects of economic of fisheries in India. *Indian J. Agril. Econ.* 28(4) : 211-218.

Rao, P.S. (1983). Fishery Economic and Management in India- Concept Publishing Company, New Delhi.

Rao, P.S. (1988). Prices and price determination for fish. Fisheries Economics and Management in India. Concept Publishing Company, New Delhi : 227-29.

Rao, P.s. and Prasad, D.N. (1978). Some price analysis of inland fisheries in Patna city market and its variation. *Sea Food Export Journal*. 10 (4) : 81-85.

Ratafia, M. and Purinton, T. (1989). Emerging aquaculture markets. *Aquaculture*. 15 (4): 32-46.

Ronsivalli, L.N. (1976). The role of fish in meeting the World's Fish needs, Massive Fisheries Reivew, National fisheries Service, Seattle, Washi, 39 (6) : 1-3.

Sarkar, S.K. (1990). Role of environmental factors on fish growth exposed to ammonium sulphate. *Fertilizers News*. 35 (3): 41-43.

Sexena, B.S. (1968). Indian fisheries in national economy. *Indian J. of Agril. Econ.* 28 (4): 219-228.

Sharma, B.K. (1978). Report of the operational Research Project on composite fish culture (1975-78). The first annual workshop on ORP of ICAR, Karnal : 15-16.

Singh, R.K.P. and Prasad, K.K. (2000). Fish Production in different Eco-system in Bihar; An Economic Analysis, *Fishing Chimes*, 20 (5) : 40-44.

Singh, R.K.P. and Singh, L.N. (1998). A Study on Economics of Fish Production in Hassanpur Block, Samastipur District (Bihar), Current and Emerging Trends in Acquaculture (ed. Thomas, P.C.),Daya Publishing House, Delhi : 78-88.

Singh, R.K.P. Singh and Prasad, K.K. (2000). Fishery Ecosystem and Fish Production in Bihar, *Acquaculture Development in India, Problems and Prospects* (ed. Krishnan, H and Brithal P.S.) National Centre for Agril. Economics and Policy Research, New Delhi-12 : 122-29.

Singh, R.K.P., Prasad K.K. and Singh, L.N. (1996). Economics of Fish Production in Hassanpur Block, Samastipur District (Bihar). *Journal of Fisheries Economics and Rural Development*. 2 (2): 1-15.

Singh, V.D. and Sampath (1984). Present status and Policy issues on Inland fisheries, Inland Fishery Resources in India (ed. Srivastava, U.K. and Vathsala, S.) Concept Publishing Company, New Delhi : 166-183.

Sinha, M. and Jha, B.C. (1997). Ecology and Fisheries of Ox-Bow
    Lakes (MAUN) of North Bihar, Central Inland Capture Fisheries
    Research Institute, Barrackpore, West Bengal.

Sinha, V.R.P. (1971). Review of composite fish culture techniques.
    First workshop on All India Coordinated Research Project on
    Composite fish culture, ICAR, Cuttack.

Sinha, V.R.P. (1975). Freshwater fish farming for more income. *Indian
    Fmg*.Vol. 28 (6&7) : 107-8.

Sinha, V.R.P. (1976). Economic evaluation of composite fish culture
    operations in different parts of India. FAO symposium on
    Development and Utilization of Inland fisheries research,
    Colombo.

Sinha, V.R.P. (1977). Composite fish culture. *Kurukshetra*. 20 (1) : 20-
    17

Sinha, V.R.P. (1978). Forest water fish farming for more Income,
    Indian Farming, 28 (7) : 107-9.

Sinha, V.R.P. and Ramachandran, V. (1985). Freshwater Fish culture-
    Publication and information division, ICAR, New Delhi.

Srivastava, U.K. (1984). Culture practices and infrastructure. Inland
    Fish marketing in India. Vol.3, Concept Publishing company,
    New Delhi.127-225.

Srivastava, U.K. (1984). Inland Fish marketing in India, Study series
    of IIM, Ahmedabad, Concept Publishing Company, New Delhi.

Srivastava, U.K. (1984). Marketing of fish. Inland fish marketing in
    India, Concept Publishing Company, New Delhi. Vol. 3 : 244-
    301.

Srivastava, U.K. (1984). Economic of fish culture. Inland fish
    marketing in India, Vol. 3, Concept Publishing Company, New
    Delhi, 202-25.

Srivastava, U.K. and Vathsala, S. 91984). Strategy for development
    of Inland fishery Resources in India – Key issue in production
    and marketing, Concept, New Delhi.

Srivastava, U.K. *et al.* (1984). Delhi transit and terminal market.
    Inland fish marketing in India, Concept Publishing Company,
    New Delhi, Vol.7 : 48-68.

Srivastava, U.K. *et al.* (1984). Economics offish culture. Inland fish marketing in India, Concept Publishing Company, New Delhi, 3: 202-25.

Srivastava, U.K. *et al.* (1984). Culture practices and infrastructure" Inland fish marketing in India, Concept Publishing Company, New Delhi, 3: 127-65.

Srivastava, U.K. *et. al.* (1984). Delhi transit and terminal market. Inland fish marketing in India, Concept Publishing Company, New Delhi, Vol. 3 : 84-93.

Suresh, P.; Selvaraj, P. and Kumar, J.V. (1888). Yield gap and constraints in Inland fish culture. *The first Indian fisheries Forum, Proceedings Assian Fisheries, Mangalore* : 439-40.

Tiwari, S.C., Negi, Y.S. and Katiha, P.K. (1994). Economics of input Resource Management in Reservoir Fisheries. *J. Inland Fish Soc. India.* 26 (1): 35-43.

Tripathi *et al.* (1983). Research highlights, low input carp culture. CIFRI, Newsletter. 6 (2) : 5.

Tripathi, (1983). Research highlights : low input carp culture. *CIFRI Newsletter* 6 (2):5.

Index

www.ingramcontent.com/pod-product-compliance
Lightning Source LLC
Chambersburg PA
CBHW050123240326
41458CB00122B/894